641 Wil
Wilson, Jason, 1974-
Godforsaken grapes : a slightly tipsy journey through the world of strange, obscure, and underappreciated wine

WITHDRAWN

GODFORSAKEN GRAPES

ALSO BY JASON WILSON

Boozehound: On the Trail of the Rare, the Obscure, and the Overrated in Spirits

GODFORSAKEN GRAPES

A SLIGHTLY TIPSY JOURNEY THROUGH THE WORLD OF STRANGE, OBSCURE, AND UNDERAPPRECIATED WINE

JASON WILSON

ABRAMS PRESS, NEW YORK

APR - - 2018
SALINE DISTRICT LIBRARY
555 N. Maple Road
Saline, MI 48176

Copyright © 2018 Jason Wilson

Cover photography copyright © 2018 Bobby Doherty

Published in 2018 by Abrams Press, an imprint of ABRAMS. All rights reserved. No portion of this book may be reproduced, stored in a retrieval system, or transmitted in any form or by any means, mechanical, electronic, photocopying, recording, or otherwise, without written permission from the publisher.

Portions of this book have been adapted from previously published material in the *Washington Post*, *AFAR*, the *Smart Set*, *Table Matters*, and *Beverage Media*.

Library of Congress Control Number: 2017949743

ISBN: 978-1-4197-2758-0
eISBN: 978-1-68335-210-5

Printed and bound in the United States
10 9 8 7 6 5 4 3 2 1

Abrams books are available at special discounts when purchased in quantity for premiums and promotions as well as fundraising or educational use. Special editions can also be created to specification. For details, contact specialsales@abramsbooks.com or the address below.

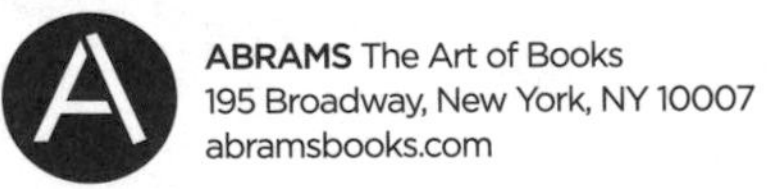

To my sons, Sander and Wes,
in hopes that they'll someday discover
their own unique tastes
(after they turn 21, of course)

A man who was fond of wine was offered some grapes at dessert after dinner. "Much obliged" said he, pushing the plate aside. "I am not accustomed to take my wine in pills."

— JEAN ANTHELME BRILLAT-SAVARIN

TABLE OF CONTENTS

The discovery of a wine is of greater moment
than the discovery of a constellation.
The universe is too full of stars.

— BENJAMIN FRANKLIN

I.

THE VINES IN THE SKY

CHAPTER 1

Dangerous Grapes

IN THE SWISS CANTON OF VALAIS, melted cheese is serious business. At the 16th-century Château de Villa in Sierre—billed as *Le Temple de la Raclette*—the evening's menu was straightforward: raclette. A guy with a long knife, called a *racleur*, scraped hot, bubbling, gooey raclette from a wheel onto warm plates that were then whisked to our wooden table, where we added small boiled potatoes served from wooden baskets, along with cornichons, pickled onions, chanterelle mushrooms, and rye bread. After that raclette, there was more raclette. For two hours, the raclette kept coming. Each plate featured a different puddle of raw-milk cheese from a different nearby mountain village. When I asked for ice water, I was gently scolded by the waiter: "Never drink cold water with raclette. The cheese will congeal into a cheese baby in your stomach."

No water was fine with me. I was at Château de Villa to drink wine with my melted cheese. And not just any wine, but wine made from some of the most obscure grapes in the world. As another round of raclette arrived, Jean-Luc Etievent, my unshaven and pastel-wearing French dining companion, poured a glass of humagne blanche. It tasted strange and big and sexy, full of ripe exotic fruit, surrounded by delicate floral aromas—sort of like mountain flowers picked by a Kardashian wearing a dirndl.

If you've never heard of humagne blanche, I don't blame you. I have been an aficionado of obscure wine and spirits for years, and I'd never heard of this white wine either. Humagne blanche dates to at least the 14th century, and in the mid–19th century it was the most widely grown grape in Valais. Now, only 75 acres of humagne blanche remain in the entire world. By comparison, cabernet sauvignon and merlot each grow on over 700,000 acres worldwide, and chardonnay grows on over 400,000 acres. With a Gallic shrug, Etievent said, "Drinking the same wines all the time is really boring."

Before I'd finished with my glass of humagne blanche, I was given a second glass by the other wine sherpa at our table, José Vouillamoz, a short, bespectacled Swiss guy in his mid-40s who wears a flat cap and kicks around his nearby hometown of Sion on a kid's scooter. "We will now taste one of the rarest wines in the world," he said, with a flourish.

Vouillamoz poured me a glass of wine made with a grape called himbertscha, which he'd helped rescue from a forgotten vineyard found high in the Alps. In the entire world, only these two acres of himbertscha exist, from which less than 800 bottles are made each year. Himbertscha is one of the strangest white wines I have ever tasted—like a forest floor of moss and dandelions that's been spritzed with lemon and Nutella. Vouillamoz took a big sip and said, "Critics claim that obscure varieties like this will never be as good as Bordeaux or Burgundy. Well, maybe not now. But what about in 50 years? One hundred years?"

We might reasonably call Etievent and Vouillamoz the Indiana Joneses of ampelography—which happens to be the study, identification, and classification of grapevines. Both are explorers on an obsessive hunt for the rarest wine grapes in the world. Vouillamoz is a world-renowned geneticist and botanist, and coauthor of the encyclopedic tome *Wine Grapes: A Complete Guide to 1,368 Vine Varieties, Including Their Origins and Flavors* (with Julia Harding and Jancis Robinson). His life's work is the study of *vitis vinifera*, the European grape species that's used to make most of the world's wine. Meanwhile, Etievent is the cofounder of Paris-based Wine Mosaic, a small nonprofit organization that works to rescue indigenous wine grapes from extinction. All over the Mediterranean, from Portugal to Lebanon, Etievent and his similarly obsessed colleagues seek out growers of rare varieties, helping farmers identify what grapes they have, then essentially serving as a support group—organizing tastings and connecting them with importers, university researchers, and wine drinkers.

I found myself in Valais because I'd grown increasingly obsessed with obscure and underappreciated wine grapes, and Etievent had invited me on a harvest-time trip to see and taste some of Wine Mosaic's most successful projects in the Alps. Here, isolated vineyards, strange microclimates, and decades spent off the traditional wine world's radar have preserved local grapes and farming traditions. In less than a decade, Wine Mosaic has saved more than 20 traditional Alpine grape varieties from dying out.

Earlier that day, about 40 kilometers from Château de Villa, Etievent and I visited the most extreme vineyards I'd ever experienced, at a craggy mountain place called Domaine de Beudon. Etievent, perhaps channeling a Parisian version of Indiana Jones, carried a pickax and wore heavy leather boots, along with royal blue pants, a white belt, and a pink scarf. We were joined by yet another rare-grape expert, Jean Rosen, vice president of a Dijon-based organization called Cépages Modestes (literally "modest grapes"). Rosen, short, stocky, and bearded, was himself a modest guy. His nickname is "Petit Verdot," after the least-known and most finicky grape used in Bordeaux blends—a variety that ripens so late that in some years the entire crop is lost. Before Petit Verdot became immersed in esoteric wine grapes, he'd been an English teacher, then an antique ceramics expert.

The only way up to Domaine de Beudon was by a creaky wooden aerial cable car—like something out of a Wes

Anderson movie. After we called up to the mountaintop on an old-fashioned phone, we waited as the cable car slowly wobbled down, and then as boxes of grapes were unloaded. A photographer traveling with us, terrified, refused to get into the cable car. Etievent, Petit Verdot, and I squeezed in, and we quickly jolted upward, suspended from a swaying cable. I could see the ground, hundreds of feet below, through the cracks between the floor and the door. About halfway up, the car lurched steeply, climbing almost vertically over a protruding rock face (the *beudon*, or "belly") that gives the winery its name. We all looked at one another wide-eyed. "Don't look down," Petit Verdot said.

We arrived at the top to fields of verbena and thyme and flowers and chickens wandering freely. The vineyards rose straight up, almost 3,000 feet above sea level. Domaine de Beudon, with its motto, *Les vignes dans le ciel* ("the vines in the sky") is considered to be one of the first and most important biodynamic wineries in the world. On the cable car platform, we met Domaine de Beudon's owner, 69-year-old Jacques Granges, who wore a bushy beard and—I kid you not—a beret. We shook hands. Granges was missing his index finger. It seemed as though we'd arrived for an audience with the mythical wizened hermit on the mountaintop.

As we sat at a table overlooking the sunny valley below, Granges brought out a dozen bottles of wine, and set down two jugs. "This one is for spitting, and this one is for dumping," he said. "I make vinegar."

"He's not going to make much vinegar today," Petit Verdot whispered to me.

Granges said little as he poured his wines. When we oohed and aahed over the first, a golden amber and chalky wine made from the chasselas grape, he said simply, "This is a wine raised by science, conscience, and a lot of love."

The next wine, from müller-thurgau grapes, was like drinking snow infused with edelweiss. "This is like magic water," said Petit Verdot. That was followed by somewhat-known sylvaner (called by the name Johannisberg in Valais) and then relatively rare petite arvine, a Swiss variety with less than 500 acres found in the world. That was followed by totally obscure reds from humagne rouge and diolinoir (each less than 300 acres worldwide).

Finally, we tasted a strange hybrid grape called chambourcin, which was created in the 19th century by crossing a French variety with a wild North American variety. Normally, a hybrid grape like this would not be permitted in a European appellation, but Granges was given special permission a few years before to plant chambourcin. "It grows in a very dangerous, steep plot," he said. "My wife wanted me to plant something there that didn't need a lot of care and attention since it's so dangerous."

I knew a number of American wineries that produced cloying, fruity, mediocre red wines from chambourcin. This mountain chambourcin was different, and for the Frenchmen with me, it was the most unusual and foreign grape of the day. "Very peculiar," said Petit Verdot as he sipped it.

As our tasting turned into drinking, fruit flies gently buzzed around crates of fresh-picked orchard fruit. My phone died, and time seemed to stop. Petit Verdot pointed toward the Great St. Bernard Pass in the distance. "This is one of the great historic places to cross the Alps," he said. "The whole region is divided into valleys. They were isolated. Historically, there wasn't a lot of communication or exchanges. You can see why each place developed its own grapes."

Even though it was brisk and cool amid the vines in the sky, all day long a bright sun shone over Valais. Finally, the sun began to set and we watched the cable car climb to meet us again.

Earlier, Granges's wife, Marion, had told us that their first cable car, years ago, derailed with Jacques inside, and he'd plunged down the mountain. He'd been badly injured and spent time in a coma. No one said a word on our descent.

A few hours later, over raclette at Château de Villa, I wanted to know: Why had grapes like humagne blanche and humagne rouge and diolinoir and himbertscha nearly disappeared?

"People became ashamed of the old-time grapes, the grapes of grandpa," Vouillamoz said. "They began planting the so-called 'noble grapes' and they would disregard the rest." *Noble* is the historic designation for grapes like chardonnay, sauvignon blanc, cabernet sauvignon, merlot, and pinot noir—the ubiquitous international grapes that made Bordeaux and Burgundy famous are now popularly grown everywhere from California to Australia to South Africa to China. "Noble grapes," Vouillamoz repeated the word with disdain. "I hate this term." What bothers people like Vouillamoz and Etievent—and me—is that while 1,368 wine grape varieties may exist, the sad truth is that 80 percent of the world's wine is produced from only 20 grapes. Many of the other 1,348 varieties face extinction.

Another raclette arrived, and it was strong and funky. Throughout dinner, I was taken by how diverse each puddle of cheese had tasted. A few were mild and creamy, one was sharp and piquant, and a couple were stinky and tangy. Much like wine grapes, I'm always surprised by how many cheeses exist in the world. As Charles de Gaulle famously said of France, "How can you govern a country which has 246 varieties of cheese?" But even de Gaulle underestimated: France has at least 400 varieties of cheese, and probably more than 1,000 if you count subvarieties. And that's just France: Hundreds of cheeses, each made according to some local tradition, exist in the rest of Europe. Those of us who quest after obscure grapes hope for a world of wine that's equally raucous and ungovernable. But wine's diversity is always under threat, and every grape that remains untasted, unknown, and underappreciated faces the risk of extinction.

After pausing only a moment to eat some cheese, Vouillamoz poured yet another rare variety, this one called gwäss. "Gouais?" said Etievent, with a raised eyebrow. Gwäss, better known as gouais blanc in French, has been banned across Europe, by various royal decrees, since the Middle Ages. That's because monarchs considered it a peasant grape that made bad wine—*gou*, in medieval French, was a derogatory term to describe something inferior. Its vines were also extremely prolific. Gwäss often took over entire vineyards, and the aristocracy didn't want a commoner mating with its noble grape varieties.

That's a curious thing I was learning about grape varieties: Each one has been created by two parents, a father and a mother, that cross-fertilize, just like you're taught in high school biology. For centuries, we could only hypothesize about a grape's parentage, but since the advent of DNA testing by scientists like Vouillamoz, we now clearly know the family tree of many grapes. Through DNA testing, for instance, gwäss has been found to be the ancient mother of around 80 varieties, several with noble pinot noir as the father, including chardonnay, gamay, and possibly riesling.

"Yeah, gwäss is kind of a slut," Vouillamoz said. His girlfriend, who was sitting next to me, shot Vouillamoz an exasperated look. "OK, OK, so we're not keeping with the times," he said. "That is a very sexist thing to say. I'm sorry. After all, we call the male grapes 'Casanovas' when they father a lot of children."

I said that it's really odd to think deeply about the sex life of grapes, especially personifying them to the point of slut-shaming. I told Vouillamoz that I doubted many people wanted to think about reproduction when they spit out an irritating grape seed.

"Yes, but they should!" said Vouillamoz. "A seed is life!"

Clearly, I'd slipped down some sort of rabbit hole into a vast alternate universe of wine geekdom.

I don't know that I've ever really emerged from that rabbit hole. The rare wines from that day at Domaine de Beudon and the dinner in Sierre loomed significantly in my mind for much of the following year. Especially one Saturday during that muggy summer week when everyone lost their minds over Pokémon GO.

All week long, instead of doing work, I'd been wandering, sweatily, around Philadelphia capturing Pokémon on my iPhone. I wasn't playing this game with my sons. No, the boys were actually away visiting their grandparents in California and I was alone, at loose ends, and I downloaded the app on my own. I found immediate, satisfying success in Pokémon's world, ignoring the reality of being a guy in his mid-40s trying not to be creepy while meandering through city neighborhoods and parks, eyes glued to the screen, flicking my finger to catch imaginary monsters. I filled my Pokédex with rare species such as the Aerodactyl, the Ponyta, the Venusaur, the Rhyhorn, and the Hitmonchan as they popped up on lawns and benches and garbage cans. After only a few days I was fast approaching Level 18. Needless to say, when I awoke Saturday morning, I was overcome with deep shame about how I'd spent my week.

Yet I couldn't help but think that Pokémon GO offered some kind of metaphor for my own life as a wine writer. Over the past couple of years, I'd spent weeks and months gallivanting around Europe, seeking out obscure wines made from rare grapes, grown in little-known regions: rotgipfler and zierfandler from Austria's Thermenregion. baga and antão vaz from Portugal, schiava or lagrein from Italian Südtirol, altesse or verdesse from France's mountainous Isère. I would sip and taste and consume those wines, then capture my impressions by jotting notes into a black Moleskine. When I thought about my life like this, it was no wonder that many friends and family members didn't consider my wine writing to be any more serious than Pokémon GO.

In any case, I decided to take a day off from Pokémon. Instead, I paid a visit to the Outer Coastal Plain wine country—which is a pretentious and boozy way of saying that I made a 35-minute drive to the semirural area of southern New Jersey near where I grew up. These days, people endeavor to make quality wine from our sandy South Jersey soil, which always invites snideness, or at least backhandedness: "The Outer Coastal Plain might be the perfect place to make fine wine in America," said the *New York Times* in 2013. "The O.C.P. has only one real challenge. It's in southern New Jersey, a state associated with many things—Springsteen, Snooki, industrial pollution, the mob—but not great wine."

People love to crack jokes when I tell them about farms in New Jersey. But Gloucester County is one of the few places where the Garden State nickname still makes sense, though even here McMansion cul-de-sacs gobble up the farmland. My family has worked in the produce business here for decades, and my cousins and I bought summer fruit for our own fruit-and-vegetable stand from the county's many farmers. Back then, the only wine I can remember in South Jersey was sickly sweet blueberry or peach wines that people bought at summer fairs.

I crossed the Walt Whitman Bridge and merged onto Route 55 right before the exit for the Deptford Mall, where I hung out as a mulleted teenager, listening to Bon Jovi and Cinderella blaring from a friend's Camaro. As I drove, I was seized by some guilt. Even though Gloucester County is not far at all from where I live and work, I rarely return for a visit. Once off the highway, I took a slightly roundabout scenic route, through Elk

Township and the community of Aura, which was once among the best peach-growing areas in the nation. As a teen, I'd learned to drive on country roads like these, steering white-knuckled next to Mr. Pickens, the gym teacher who taught Drivers Ed. wearing polyester coach shorts and a whistle around his neck. In Aura, I grew a little sad when I didn't see many fruit trees. There was, however, a large housing development called The Orchards. Soon enough, things got more rural, as I turned on to Whig Lane, then Elk Road past the Hardingville Bible Church and Old Man's Creek Campground. I passed a Christmas tree farm, a used tractor-trailer cab for sale in a front yard, and finally some apple and peach trees. I thought about filling a bucket with some berries or peaches at a U-pick spot called Mood's Farm, as my family has for years. But this day happened to be Mood's Farm's blueberry festival and the place was teeming with crowds eating blueberry pie and blueberry ice cream and drinking blueberry cider. So I just bought an apple cider doughnut and moved on.

Mood's was just down the road from Heritage Vineyards in Mullica Hill, where I intended to taste wine. When I pulled up at Heritage, there was a guy playing acoustic guitar and singing outside on the patio. Outside, it looked like the kind of place where you'd pick pumpkins or go for a hayride. Inside, the tasting room looked like a country store, with wine tchotkes and knickknacks for sale, including some decorative signs that would not have been out of place on a Jersey Shore boardwalk: "A Meal Without Wine Is Called Breakfast"; "Today's Forecast 100% Chance Of Wine"; "I Just Rescued Some Wine. It Was Trapped in a Bottle."

Heritage Vineyards gained some renown a few years earlier at a blind tasting hosted at Princeton University during the annual conference of the American Association of Wine Economists. This so-called Judgment of Princeton pitted New Jersey wines against those from Bordeaux and Burgundy, including top châteaux such as Mouton-Rothschild and Joseph Drouhin. It was closely modeled after the famed 1976 Judgment of Paris tasting in which top critics unknowingly selected California wines over French wines, a heresy at the time that helped establish Napa Valley on the global wine map. George Taber, the same journalist who'd originally reported about the 1976 Paris tasting for *Time* magazine, was actually the moderator in Princeton, where nine wine experts from the United States, France, and Belgium convened to judge the wines.

In keeping with the now-expected "gotcha!" nature of these blind-tasting events, the wine experts gave essentially the same scores to the New Jersey and French wines. The usual controversy ensued. After the identity of the wines was revealed, at least one judge, Odette Kahn, editor of *La Revue Du Vin De France*, demanded her scorecard back. She had rated New Jersey wines as her number one and two choices. Meanwhile, local news outlets had a hearty populist chuckle at the expense of the Francophile wine snobs.

The Judgment of Princeton was a boon to Heritage Vineyards. Its 2010 BDX Bordeaux-style blend won third place among the reds, scoring just a half-point behind Haut-Brion, the famed Bordeaux château. Heritage's 2010 chardonnay also finished third, ahead of a few bottlings of Montrachet, the Burgundy grand cru. I've tasted both of these Heritage wines and they're very accomplished, and delicious—and at $50, the Bordeaux-style red blend is at least $500 to $1,000 less than the Château Haut-Brion. Still, I've often pondered why up-and-coming wineries and regions still look to Bordeaux and Burgundy as their benchmark. And it's not just in South Jersey. No matter what wine region one visits—Chile, Australia, Oregon—so many winemakers still aspire to craft wines from cabernet sauvignon and merlot and chardonnay and pinot noir grapes, and so many wine lovers aspire to drink these wines.

There is, of course, nothing intrinsically wrong with trying to replicate the prestige of cabernet sauvignon, chardonnay, or pinot noir. Unless you agree with Jean-Luc Etievent that "drinking the same wines all the time is really boring." Or perhaps you're like me, and you're excited about having new experiences and learning new things. Every new grape you've never tasted before, after all, offers the chance to taste a new flavor. In this globalized world, more and more of us seek out the local specialties. These grapes carry with them a taste of place and culture. We endeavor to preserve these grapes, then, for the same reason we save heirloom tomatoes and apples and heritage cattle, and build vast seed banks. Within these organisms may lie clues to solving the challenges of climate and disease, as well as recording the historical record of human taste.

I know I'm not alone. In recent years, there's been a revived interest in little-known grapes from lesser-known regions. For instance, two decades ago, grapes like carmenère from Chile, grüner veltliner from Austria, or albariño from Spanish Galicia were completely off the radar. Now they're old hat to many wine enthusiasts. So the quest for even rarer grapes has intensified. While the new generation of sommeliers in the hip, big-city wine bars can often be overbearing, with their disdain for any wine that might be considered mainstream, we can thank them for creating a demand, however small and exclusive, for rare grapes. Often, a sommelier's love of obscure wines might seem like an embrace of obscurity for obscurity's sake. But sometimes, it can also serve a higher purpose.

When I arrived at Heritage, ready for a world-class tasting, the woman pouring wines asked, "Sweet or dry?" When I looked around the bar, I was surprised to see many people drinking fruit wines that read Jersey Blue or Jersey Sugar Plum. The pour woman looked relieved when I said "dry" and opted for both the Classic and Reserve tastings.

While I was looking forward to tasting their acclaimed Bordeaux-style wines, I was quickly enamored by how Heritage used the lesser-known Bordeaux grapes. For instance, when I tasted the 2014 sauvignon blanc I commented that it had a fleshy orchard-fruit element that wasn't typical. The pourer told me that about a quarter of the wine was made from the sémillon grape. This made sense, since sémillon is a classic blending partner with sauvignon blanc in dry Bordeaux whites, as well as sweet Sauternes. But then the pourer swiftly opened a bottle of 100 percent sémillon. "We originally planted this for blending, but it was just so good," she said. I agreed; this was a surprising, uncommon wine, something you rarely see, even in Bordeaux. Sémillon is "not a fashionable variety," according to Vouillamoz and his coauthors of *Wine Grapes*. Worldwide, there is about five times more sauvignon blanc, and about eight times more chardonnay grown than sémillon. That may be because in much of the world, sémillon can be too fat and honeyed and cloying. In fact, the only other place in the world you drink an exceptional dry 100 percent sémillon like this is Australia's Hunter Valley (one of the "wine world's enigmas" according to wine writer Oz Clarke). In any case, Heritage's Jersey rendition of the grape was fresh and lithe and elegant. The unrepentant wine geek in me started to get tingles of excitement.

A similar thing happened when I tasted Heritage's Bordeaux-style 2011 Estate Reserve BDX, which was a blend of 40 percent cabernet sauvignon, 32 percent merlot, 16 percent petit verdot, 8 percent cabernet franc, and 4 percent malbec. Maybe I was subconsciously toasting Petit Verdot, my drinking companion that day at Domaine de Beudon, but what struck me was the high percentage of actual petit verdot in the bottle. Yes, this all may seem like the pinnacle of wine geekiness: to single out an obscure grape like sémillon or petit verdot. But please stick with me. Since petit verdot is known for its rustic character, dark color, powerful flavor, and strong tannins, it's traditionally used like a chef would use seasoning. Few châteaux in Bordeaux ever use more than around 3 percent—16 percent is off the charts.

So Heritage's Estate Reserve BDX boasted quadruple or quintuple the typical seasoning. Why? Perhaps petit verdot grows much better in South Jersey's soil and climate than it does in Bordeaux, where the grape only reaches full ripeness about once out of every four harvests. Perhaps South Jersey's petit verdot ripens more fully and therefore has more complex fruit, softer tannins, and maybe some of the coarse rustic character is tamed. Perhaps Jersey petit verdot is, dare we say, more elegant than Bordeaux petit verdot? In any case, petit verdot—an oddball grape by any measure—makes a big difference in this very un-Bordeaux Bordeaux blend.

Several years before, I'd actually written one of those typically backhanded newspaper articles about Outer Coastal Plain wine. (Yes, I'll admit, I, too, made a Snooki reference.) At the time, one winemaker told me, "You can make a good wine anywhere. You just have to pick the right grape." This, of course, is the fateful decision that every winemaker in every wine region in the world must grapple with. If you happen to be the great-great-great-grandson of a Bordelais winemaker who several centuries ago figured out that cabernet sauvignon vines grew really well on that hill yonder, well then you're all set. You just tend the grapes your grandfather planted and don't screw it up. Château Haut-Brion can document grapes being grown as early as 1423. Heritage, on the other hand, planted its vines in

1999. It's almost 600 years behind Bordeaux in figuring out what works and what doesn't.

As I tasted the next wine at Heritage, their 2013 cabernet franc, I regretted being so skeptical. I am now convinced that cabernet franc should definitely be South Jersey's grape. My thinking may be prejudiced in part because I happen to love cabernet franc. For years, I've thought, if you are becoming more adventurous in your wine drinking, if you're curious to explore beyond plump, ripe, and oaky, if you like eating and drinking food and wine together at the same time . . . well then you really should try drinking more cabernet franc. In Paris, Loire Valley reds made from 100 percent cabernet franc have been traditional house reds in bistros for decades, underscoring how well it pairs with so many different dishes. Why cabernet franc remains so unpopular in the United States, however, has perplexed me. Just like cabernet sauvignon and merlot, it's one the official grapes of Bordeaux blends. In fact, cabernet franc is the parent of cabernet sauvignon, so its origin is even more ancient. Ampelographers consider it to be a so-called "founder" grape. Nowadays, though, it's mostly an afterthought, and cabernet franc lags far behind its offspring—it's only the 32nd most planted grape in the world, with less than a fifth of cabernet sauvignon's acreage.

One issue is that wines from cabernet franc's spiritual home, the Loire Valley, can be very challenging and, well, very French. Give most American consumers a label that reads Chinon or Bourgueil or Saumur-Champigny and their eyes will glaze over. "Are those medical conditions or characters on *Game of Thrones*?" someone once asked me. No, I said, they're places in Loire Valley which produce cabernet franc wines. Beyond nomenclature, these wines are known more for their more herbal, savory flavors and aromas, usually with little or no oak barreling. The fruit is subtle; Loire Valley reds don't loudly announce themselves or smack you over the head. This, of course, is in contrast to the big, fruity wines that plenty of people still favor. Critic Michael Steinberger once wrote about cabernet franc wines: "People accustomed to plusher California wines often find them too austere . . . in the minds of many American consumers, synonymous with words like *thin* and *weedy*." Sadly, those consumers are missing out on some of the most drinkable wines on the planet—*drinkable* being, in my book, just about the highest virtue of a wine.

An appreciation of savory wines is slowly spreading among younger wine drinkers. Maybe cabernet franc isn't a wine that finance bros, neckties thrown over their shoulders, will expense at a steakhouse. But if you haven't been paying attention, here's a news flash: People are eating fewer warm-blooded animals these days, and so wines have to pair with dishes besides medium-rare meat. While cabernet franc can evoke "green"—olive, pepper, tobacco leaf—there are also plenty of fresh berry notes and the lighter body and unique underlying sensations of graphite, iron, or even pencil shavings. (Again, stick with me; this can be a good thing.) Perhaps surprisingly, a wine with notes of pencil shavings and olive match really well with the foods we actually eat these days.

In any case, the cabernet franc at Heritage, with its notes of dark fruits and thyme and autumnal mulled spices, was special without trying to be special. It was a little softer and plumper than one from the Loire Valley, but still attractive and desirable—a cabernet franc with a dad bod, a comfy plaid flannel, and a beard.

I already was feeling really warm and fuzzy in the tasting room when I saw the final wine on the menu. It was Heritage's 2013 chambourcin, which was the same wine that Jacques Granges had planted in that steep, dangerous vineyard at Domaine de Beudon. A decade ago, when wine in New Jersey was still a novelty, a number of local winemakers envisioned chambourcin, with its hybrid of European and wild North American DNA, as the state's calling-card grape. But the wines produced from chambourcin are strange, often with a so-called "foxy" or wet-fur aroma associated with North American grape varieties. Heritage's inky-purple chambourcin avoided the foxy smell, but it did have a curious nose of Dr Pepper and roasted cashews and tasted of blackberry jam on raisin bread. As Petit Verdot had said of the chambourcin at Domaine de Beudon: Very peculiar. But oddly pleasant.

When I finished my tasting, I paid $25 each for the bottles of the cabernet franc and the chambourcin, and ordered a small presliced meat-and-cheese plate that came wrapped in plastic. I took everything, along with two wine glasses, out to the patio, where the middle-aged guy with the guitar strummed and sang. I unwrapped the plate, and nibbled on sweet soppressata, pepperoni, and prosciutto (or pro-ZHOOT as they say in South Jersey), as well as a hunk of Prima Donna, a popular cheese that's a hybrid of Parmigiano Reggiano and aged Dutch gouda. I uncorked and poured both wines—an older couple sitting at the next table saw this, and raised their eyebrows. I had no way of telling them that I was a professional and would only be taking a few sips and bringing the rest home. They likely wouldn't have believed me anyway. So I just let them assume I was a decadent drunk settling in for a solitary two-bottle Saturday afternoon.

I swirled and tasted a little more cabernet franc, then the chambourcin. As if on cue, the musician broke into a rendition of Bruce Springsteen's classic "The River." As a native son, I know the Boss's words by heart. Once, in a whiskey bar in Edinburgh, Scotland, I watched a bumbling singer lose his place and forget the lyrics while performing "The River" live. Immediately, I raised my voice, chiming in to finish the song—as though I'd prepared for that moment my whole life.

Listening to "The River" in the middle of Gloucester County, while drinking local cabernet franc and chambourcin, I suddenly began to feel a profound sense of what wine people call terroir. I'd never thought about a South Jersey terroir before, and now that I did, there was nothing pretentious about the idea at all. Why couldn't New Jersey be the next Bordeaux or Napa Valley? But there was more to it than that. Drinking these uncommon South Jersey wines made me feel connected to the larger world of wine. The chambourcin, in particular, made me think of my visit to Valais and to Domaine de Beudon, where Jacques Granges, bearded and bereted, worked the mountaintop vineyards.

As I sipped my wine, I grabbed my iPhone. Instead of opening Pokémon GO, I typed "Jacques Granges Domaine de Beudon" into Google. Suddenly, I was shocked to see the headline that popped up: *Jacques Granges, pionnier de la biodynamie, meurt dans un accident*. I don't speak French but I didn't need a translation of *accident*, and I knew *meurt* meant he had died. The article had been published in a Swiss newspaper three weeks before. Apparently, Granges had been working in a steep vineyard when his tractor overturned, tragically crushing and killing him.

In that moment, my mind left New Jersey, drifting above as if on cable car toward the vines in the sky. Had Granges been working the dangerous chambourcin vineyard? Who would now carry on the work of cultivating petite arvine and diolinoir and humagne rouge? Even though I'd only spent two hours with the man and didn't really know him at all, I was distraught in the same way as when I learn that an artist I admire has suddenly died. I thought about what Jean-Luc Etievent, with his pickax and pink scarf, had said that day in Domaine de Beudon: "These sorts of vineyards are treasures. There's a huge necessity to protect them. Wine can be quaint and romantic, but in the end it's a business. And you can die in this business."

The middle-aged Springsteen impersonator, wailing on an out-of-tune harmonica, segued from "The River" to "Atlantic City" and I was almost physically snapped back to South Jersey. A big white SUV limousine pulled up and a shrieking bachelorette party streamed out. I knew it was time to leave. I finished my last pieces of pepperoni and Prima Donna, took a last swig of chambourcin, and made one last silent toast to Jacques Granges. Then I corked my wines, walked to my car, and drove away.

CHAPTER TWO

Château du Blah Blah Blah

IF I WANTED TO EXPERIENCE as many of the 1,368 grape varieties as I could, I knew I'd need to expand my wine tasting horizons and venture away from my South Jersey terroir for a while. So, several weeks later, I prepared to travel to Europe again.

On the September morning I was due to depart, I found myself eating fresh concord grapes at breakfast with my son Wes, who was then in elementary school. We'd picked these concord grapes near Heritage Vineyards, at Mood's Farm, where they are only available for a few weeks in late August and early September. Eating fresh, late-summer concord grapes that you've picked yourself is a wonderful experience, totally different than eating industrial seedless grapes that have been shipped in from Chile.

Wes was trying to become lieutenant of the fifth grade school safety patrol, and he had to give a speech that day to convince his fellow safeties to vote for him. I'd just asked him to rewrite his speech because he'd spelled lieutenant wrong once (*lutendant*) and responsible wrong twice (*risponsable*) and in some spots he couldn't even read his own handwriting. So, exasperated, he scribbled away at our kitchen table.

Wes suddenly looked up from rewriting his speech. "Do they make wine out of these grapes?" he asked. For my son, this was not an unreasonable question. He'd been exposed to wine from an early age, witnessing his father conduct tastings at this same kitchen table, and having traveled with me to vineyards and wineries in Spain, Italy, and Austria.

"No," I said. "Not usually this kind of grape. This is a grape they use for juice or jelly."

He seemed disappointed. And then irritated as he erased another misspelling of *risponsable*. "Why not?" he asked.

"Well," I said. "Some people try to make wine out of them. But the wine from concord grapes is usually really bad."

"But these taste so sweet and so good," he said. "Why do they make a bad wine?"

My head began to swim with too much information: The concord grape was discovered in 1840, growing wild along the Concord River in Massachusetts. Concord is believed to be the offspring of a native North American species called *vitis labrusca*, which is completely different from European *vitis vinifera*, which is much better suited for winemaking. Wines made from *vitis labrusca*—literally the "fox grape"—usually have a sharp, off-putting "foxy" aroma. Concord grapes are a slip-skin variety, meaning the skins will separate too easily from the fruit, making pressing difficult. Concord grapes are generally too high in acidity to be palatable and too low in sugar to achieve the alcohol levels necessary for wine.

None of this, of course, would make much sense to Wes. So I said: "They just do. They just make bad wine."

"Well," he said. "Do you think maybe people just haven't figured out how to make a good wine from them?" He put down his pencil. Wes may as well have been a police lieutenant continuing an interrogation.

"What level of sommelier are you?" he asked me.

"What?" I said. "What do you mean?"

"We watched that movie about those guys studying to be master sommeliers," he said. I realized he was talking about *SOMM*—the documentary that suddenly made being a wine waiter hip and drove lots of young people to seek certification as a sommelier. Wes had recently watched it along with me on the couch. "So what level do you have?"

"I'm not a sommelier," I said. "I don't have any level of certification."

"Oh," he said with a shrug, and went back to writing his speech. I ate another handful of concord grapes in silence. My son had unwittingly and keenly broached a sensitive topic. I was being authoritative, or perhaps even authoritarian, in my declaration on the poor quality of concord grapes. But who was I to dismiss wines made with concord, or any other grape with *vitis labrusca* parentage, or any wine made from any grape in the world?

The issue of authority and credentialism crops up again and again in the wine business. Wine is lousy with certifications, dogmas, designations, gatekeepers. I move through this world—sometimes awkwardly—without a framed diploma on my wall declaring me an expert. I am not a famous or powerful wine writer, such as Robert Parker, still the world's best-known and most influential critic. I am not someone with a lapel pin from the Court of Master Sommeliers. And I am certainly not wealthy enough to be a collector who bids on great vintages at wine auctions.

While I eventually started being paid to write about wine, all my knowledge has come from the seat of my pants, following my passion, traveling and tasting where I could afford—and often where I could not. This is not to say that I don't have a memory library of thousands of wines, along with in-depth travel to many wine regions, and visits with many winemakers, in my head. But it's been a wildcat education, unstructured and self-taught. One of my personal mantras has been a quote from Flaubert's *Sentimental Education*: "Exuberance is better than taste."

Still, anyone who's ever taken even a single wine class has been presented with the idea that wine is an aspirational ladder you must work hard to climb, and at the top of the ladder are the so-called Serious Wines, such as Bordeaux and Burgundy.

I remember facing this ladder my first time in Bordeaux, several years before. This was after I'd begun publishing regular wine articles, mostly on up-and-coming Italian and Spanish regions, and I decided I needed to make a proper pilgrimage to Bordeaux. I'd begun chasing the topic of wine in earnest after spending more than six years reporting on, and eventually writing a book on, spirits and cocktails. But I'd grown bored of discussions over barrel-aged Negronis and house-made bitters and artisan mezcal and techniques on carving proper ice cubes. I figured my drinks knowledge would make for an easy transition to wine. I had no idea how wrong I was.

Wineries in Bordeaux are still ranked by a classification system demanded by Emperor Napoleon III in 1855 for the Exposition Universelle de Paris. The prestigious *premier cru* wineries that fetch the highest prices today are the same ones that had prestige and fetched high prices during the Second French Empire 160 years ago. Prior to that, Bordeaux had been the prestige wine exported in the 17th century to Holland during the Dutch Golden Age. And before that, Bordeaux had been exported to Britain since at least the 12th-century wedding of Henry II and Eleanor of Aquitaine. So long story short, Bordeaux has always been about power and money.

This all hit me when I found myself sitting uneasily in the tasting parlor of Château La Mission Haut-Brion in the company of Prince Robert de Luxembourg, the château's royal managing director. Prince Robert told me that the big-time critics like Robert Parker had visited here the week before. During our chitchat, I mentioned this was my first trip to Bordeaux, and the Prince guffawed incredulously. "Never been to Bordeaux? And you write about wine?"

"Um, well . . . yeah?" I said, immediately backpedaling. "I guess I've just spent most of my time in places like Italy and Spain and Portugal. And other parts of France? I don't know. Italy, I guess, is where most of my wine knowledge has come from."

"Oh," said the Prince, in a grand princely fashion, "so you are an expert in Italian wines? Ha. Well, we have an Italian wine expert here!" I haven't felt so foolish since middle school when I forgot to wear shorts to a basketball game, and pulled down my sweatpants to reveal my tighty-whities to the crowd. The message from Prince Robert seemed to be: How the hell did you get an appointment to taste wines with me?

I looked around at the regal tasting room, with the heavy wood furniture and the bust of someone presumably famous, and the high-seated chairs where the important wine critics swirl and spit and opine and move cases of thousand-dollar wine. And I decided to jump right in with a question that may have been impolite: "A lot of wine writers and sommeliers back in the States say that Bordeaux isn't really relevant anymore. What do you say to those people?"

"The fact is," said Prince Robert, "that people need to write about something. And Bordeaux is obviously so relevant that they need to write something about Bordeaux. It's the tall poppy syndrome."

Prince Robert clearly had answered this question many times before. "I would ask other winemakers around the world and they will tell you that Bordeaux would be the benchmark by which to judge all other wines," he said. "There are no wines in the world that receive more excitement."

"But wait," I said. "Aren't you worried that younger people are not drinking Bordeaux? That it's not even on their radar?

Aren't you afraid that when this generation can finally afford your wines, they won't care about them?"

"Yes, the young wine drinker likes the simplicity of New World wines. Wines that are easy to explain," he said, and I'm not sure I can properly convey just how much contempt dripped from the Prince's voice, contempt seemingly for an entire generation of young wine drinkers who could not yet afford his wines.

"Anyway," he said, "I am confident that people will come back to the great wines of Bordeaux. There has never been more demand for the top-end wines." This may be true, but the market for Bordeaux wines is now being driven, in large part, by newer collectors in Asia. One might reasonably hypothesize that tastes will eventually change in China and India, too, just as they have in the United States in the decades since the 1980s, when Americans "discovered" Bordeaux (via Robert Parker). Surely by now there is a Chinese Robert Parker? And in the not-so-distant future a backlash against Bordeaux by young, tattooed, hipster Chinese sommeliers will happen?

I didn't get to ask these questions because, apparently, our conversation bored the Prince. He rose from his chair, bid me adieu, and wished me a good first trip to Bordeaux. "Enjoy those Italian wines," he said, with a smile and a wink.

I was then left to taste nine wines from the 2011 vintage with the public relations person. How were the wines? Amazing. No doubt about it. The flagship first label wine was as complex and dense and rich as just about anything I'd tasted at that point in my wine education. But at what price? Château Haut-Brion is regularly listed at over $1,000 a bottle. I tasted the only ounce of the 2011 that I will likely ever taste, one ounce more than most of my friends and readers will likely ever taste. Will my description inspire you to drink Bordeaux? I mean, one of my friends drove a Ferrari once and another had a date with a Victoria's

Secret model, but neither of their descriptions has exactly led me closer to the same experience.

Is there any wine that intimidates more than Bordeaux? Even among friends of mine who are Serious Wine drinkers, it feels like the schoolyard bully that no one wants to stand up to. "I am totally, totally intimidated by Bordeaux," sheepishly admitted one acquaintance, a knowledgeable food writer who'd already passed the first level of the sommelier certification. "I walk past that shelf in the store and all the Bordeaux bottles look exactly the same. Same colors, same script-y fonts, same gold leaf, same illustration of the damn château. It's always Château du Something Something. Château du Blah Blah Blah. Château du Frenchy French. How do I even know where to begin?"

When I once raised the topic of Bordeaux with another friend, a beverage manager at a fine restaurant, he got seriously angry. "Ugh, why do I even care about Bordeaux?" he nearly shouted. "Who is able to afford it? Why don't they just sell it all to Chinese billionaires?"

I feel their exasperation. Every year, we hear breathless hype on Bordeaux, from astronomic auction prices, to the Talmud-like annual spring dispatches from Bordeaux *en primeur* barrel tastings, to the crystal-ball predictions on futures (i.e., wine bought by rich people that hasn't even been bottled yet). The 2009 vintage was especially full of hyperbole. Parker, for instance, wrote: "For some Médocs and Graves, 2009 may turn out to be the finest vintage I have tasted in 32 years of covering Bordeaux." Yet I feel like I hear a declaration to this effect from one or two of the major wine critics at least every other year. In fact, the very next year, 2010, was similarly declared the "vintage of the century" by numerous critics.

Before my trip to Bordeaux, *Wine Spectator*, the industry bible, devoted its March 2012 cover to Bordeaux's 2009 vintage, which the magazine deemed "classic." A cover line declared

"Second Labels Offer Great Value" and underneath—presumably to show this so-called "great value"—were images of labels from Les Forts de Latour and Carruades de Lafite, both with the standard, staid Bordeaux pen-and-ink illustrations of the estates. Next to the labels, editors listed the prices and critics' scores right on the cover: $345 for the 93-point Les Forts de Latour and $400 for the 92-point Carruades de Lafite.

Now, I'm pretty certain that irony was not at play, since I doubt there is a less ironic magazine in the world than *Wine Spectator*. I'm therefore going to assume that the editors earnestly believe that paying $400 for a bottle of wine represents "great value." Perhaps these editors would say the word "value" is relative. The best wines, the premier cru "first labels," were also featured on the magazine cover as a point of comparison next to their "second labels"—Château Latour ($1,600; 99 points) and Château Lafite Rothschild ($1,800; 98 points). But just to be clear: For $400, you won't be getting the winery's best grape juice. For $400, you'll be buying a wine that, by the magazine's own numeral ratings, essentially received an A-minus.

I learned a lot from my trip to Bordeaux. After all, it's a vast wine region of 54 different appellations and 8,500 producers, who make wine at all price points, and about half of it sells for less than $7 per bottle. But it was disquieting to see how much emphasis was placed on the premier cru wines. It also seemed a little disquieting to some of the younger winemakers I met.

At Château Pape Clément, which had its first harvest in 1252, I met with a 27-year-old assistant winemaker named Arnaud Lasisz. "In Bordeaux," he told me, "everyone always says we have to be the example. We cannot make middling wines."

As we wandered the cellars, Lasisz told me, "Our boss, he got a 95 for the red and he called a meeting and was furious. He said to us, 'Tell me what you need, and I will give you the money to do it.' He was furious. Even though he'd gotten 100 for the white that year!"

"So money is no object, right?" I said. "Don't you have the resources to do what you want?"

"It's not as simple as that."

Before Lasisz came home to Bordeaux, he'd previously worked in New Zealand and Australia. "In Bordeaux, there are too many rules sometimes. I was surprised in New Zealand and Australia by all the experimentation. There are no rules."

As we tasted the peculiar, intense 1988 vintage, full of cigar smoke and tar and leather and dried leaves, I asked Lasisz—as a guy in his 20s—if he drank a lot of Bordeaux wines when he's out at night with his buddies. "When I'm with friends who aren't in the industry, they want something different," he said.

They're not the only ones who want something different. Since my own encounter with Bordeaux, I've read a steady stream of wine writers confess that the world's benchmark wines don't really excite them anymore. Several years ago, Matt Kramer, a top columnist at *Wine Spectator*, made one of the first public confessions. "Today, more than ever before, I find myself with an almost urgent taste for emotions of surprise," Kramer wrote in a piece titled "Why I No Longer Buy Expensive Wine." He called out expensive wines as "utterly predictable" and "rarely offer that element of surprise for me anymore."

As prices have risen further, this sentiment gained steam. In August 2016, the respected British wine critic Andrew Jefford wrote in *Decanter*, "At some point over the last year, I realized something had changed in my relationship with wine. I didn't want the best any more . . . It may be exquisite, but it isn't necessarily interesting."

Just a few weeks later, Jon Bonné, in the *Washington Post*, expressed his own disillusionment with Bordeaux and

Burgundy—as well as in Napa Valley, Tuscany, and elsewhere. "I've come to an awkward conclusion," Bonné wrote. "Yes, those wines are great. But I can live without them." He added: "The great, to crib a line from Voltaire, has become the enemy of the good."

Wine books almost always begin with a light-hearted tale of the author's initiation into the world of wine via some crappy bottle of plonk. This is where you'll normally read an anecdote of misguided youth involving, say, Thunderbird, Sutter Home white zinfandel, Boone's Farm, Lancer's, Mateus, Korbel, Bartles & Jaymes wine coolers or—for the generation of wine books still to be written by younger millennials—boxes of Franzia. It's sort of like an immutable law, so who am I to violate it?

My own wine story begins during my senior year of high school when I was very enthusiastic about Mogen David's flavored and fortified wine MD 20/20, otherwise known as "Mad Dog." MD 20/20's Orange Jubilee was my particular tipple of choice, and the reason had more to do with how much easier it was to hide in the woods than a six-pack of beer. I vaguely remember it tasting like a mix of chalky, watered-down Sunny D and grain alcohol, but I've mostly tried to cleanse that memory from my mind, along with numerous other suburban public school rites of passage.

My MD 20/20 connoisseurship ended soon after I left for college in Boston. During the first week of college I professed my enthusiasm for Mad Dog and shared some Orange Jubilee with the new friends on my dorm floor. After gagging and spitting out the MD 20/20, my new friends laughed and gave me the ironic nickname "Mad Dog," which stuck until I transferred to the University of Vermont after my freshman year. It was an early lesson in how fraught it can be to express a wine preference, as well as a lesson in how it feels it to have one's taste disapprovingly assessed.

In reality, there was no reason my first "wine" had to be MD 20/20 Orange Jubilee. My father was of the generation that, in the late 1970s and 1980s, leaped headlong into an appreciation of Napa and Sonoma cabernet sauvignon and chardonnay—all aspiring to compete with Bordeaux and Burgundy. There were often bottles of Kendall-Jackson or Robert Mondavi or Grgich Hills or Beringer opened at dinners and parties. I occasionally had a taste, but back then I had little interest in drinking what my parents drank.

So it wouldn't be until the summer after my sophomore year, when I was 19, that I first truly experienced wine. I was studying abroad in Italy, living with a family in a village called Pieve San Giacomo near the Po River in the province of Cremona. Every night, Paolo, the father, sliced a plateful of prosciutto and cut a hunk from a wheel of Grana Padano. Then he uncorked and poured a fizzy red, chilled, from an unlabeled liter bottle he'd fetched from a dark corner of the barn—the same barn I'd wandered into one morning and saw him butchering a cow. Paolo didn't go for fancy wine glasses, but rather used what we would have called "juice" glasses back home in Jersey. Beyond preparing sliced meat, cheese, and wine, men were otherwise forbidden in his wife's kitchen, so while Anna busily made us dinner, Paolo and I would sip our cool, fizzy red wine from juice glasses in front of the television, blaring a soccer game on those hot evenings.

I had never tasted or witnessed a wine like this. The liquid was bright purple, with a thick pink foam that formed as it was poured. I knew enough to know that the Napa cabs on my parents' table back home didn't foam. Paolo's wine certainly tasted fruity, though it was more tangy than sweet, and what made it foreign to me was the aroma. My father's wines smelled like identifiable fruits—plums, cherries, berries—unlike this fizzy wine. It was a little stinky, to be honest, but in a very pleasant way. Sort of like the beautiful hippie girls I had crushes on back at college in Vermont. I didn't have the language back then, but in my memory the aroma smelled earthy, rustic, fertile, alive, almost like the essence of the farm and dusty streets of the village. Back then, it simply smelled and tasted like the Old Europe I had hoped to find.

Of course, being young and naïve, I never bothered to ask Paolo anything about his wine—the grapes, where it was made, who made it. I kept in touch with the family, but since Paolo died in the late 1990s, and since neither Anna nor his daughter Daniela drink wine, the fizzy red's provenance remained a mystery. Over the years, though, as my wine knowledge grew, I hypothesized that what I'd been imbibing those summer evenings long ago had been lambrusco, mainly since Pieve San Giacomo is just over an hour's drive from Modena, lambrusco's place of origin.

As I moved further into drinking and learning and eventually writing about wine, I occasionally told Wine People I met at trade tastings and industry events about enjoying this fizzy red wine as a 19-year-old, and it never failed to draw a chuckle. "Lambrusco!" they'd say. "Riunite!" For decades, cheap and sweet Riunite lambrusco had been one of the best selling wines in the United States. During its heyday in the early 1980s, I can remember seeing those cheesy "Riunite on ice. That's nice!" commercials when the babysitter let us stay up late to watch *The Love Boat* and *Fantasy Island*. But as Americans' knowledge increased during the 1990s, budding wine connoisseurs didn't want to hear about fizzy red wine anymore.

So even though the stuff I drank back in Pieve San Giacomo was neither sweet nor tacky, I just stopped talking about it, or even thinking about it. Like so many other aspirational

Wine People of my generation, I dutifully learned to appreciate Serious Wines, which in the late 20th and early 21st centuries still mainly meant cabernet sauvignon and pinot noir and chardonnay from various pricey bottlings. Instead of rustic Italian wine, I delved into high-end Barolo, Brunello di Montalcino, and Super Tuscans. There seemed to be a general consensus on what defined a Serious Wine. "Frankly, there are no secrets about the origin and production of the world's finest wines," wrote Robert Parker in an essay called "What Constitutes a Great Wine?" in *The Ultimate Wine Companion*. Serious Wines, according to Parker, pleased the intellect as well as the palate, offered intense flavors and aromas, and always improved with lengthy aging. I've read similar things from other critics. When talking about Serious Wines, very little is said about drinkability or refreshment.

So I filed away my old "unserious" fizzy red into a similar place as my youthful Orange Jubilee. I was being schooled by wine educators and sommeliers and wine critics: As a *sophisticated* drinker, a Wine Person, I was made to believe I should be moving beyond things like fizzy reds; I should be climbing the ladder, constantly reaching upward, leaving behind so-called *lesser* wines and striving toward greatness, toward the profound, and toward—inevitably—expensive Serious Wines.

Two decades after my summer in Pieve San Giacomo, I found myself in Italy's Langhe region, in Piedmont, visiting a bunch of producers of Barolo, the complex, elegant wine made from the nebbiolo grape—the epitome of a Serious Wine. It was the court wine of the House of Savoy monarchs in Turin, gaining its nickname as "the wine of kings, the king of wines." I tasted dozens of astonishing and often profound and transcendent Barolos—some of which had been aged for decades—which convinced me, once again, that nebbiolo grown in this corner of northwestern Italy creates one of the world's greatest wines.

My visit culminated on a sunny Sunday afternoon, when I attended an auction called Asta del Barolo. On the day of the auction, I climbed the village of Barolo's winding, narrow cobblestone streets up to a castle overlooking miles and miles of some of the most valuable vineyards in the world. Bottles from prized vintages sold to collectors—some from as far away as Shanghai, Moscow, and Dubai—for thousands of dollars. A group of men bid live via video chat from a restaurant table in Singapore. One acquaintance, an Austrian banker living in Hong Kong, paid 3,000 euros for three magnums dating from the mid-1980s. I sat next to a charming producer, whose family's elegant, silky, Barolos annually receive high scores from critics, who call them "genius" and "breathtaking." During the many courses of lunch, we tasted about 15 examples of the 2009 vintage. Later, there was talk among the younger winemakers about Jay-Z and Beyoncé's recent visit to Barolo for a weekend, where they supposedly dropped $50,000 on wine and truffles.

I won't lie. It is seductive to be part of an afternoon crowd like that. And I cannot state clearly enough how much I do enjoy Barolo. Perhaps it is nerdy to say, but it can be like listening to a beautiful, challenging piece of music or standing before a grand, moving work of art. I love it so much that when people ask what my favorite wine is, I often exclaim, "Barolo!" And they nod, and say, "Ah, yes. Barolo, of course."

But that afternoon at the castle was total fantasyland. When I returned home, would I be drinking very much Barolo? No, not so much. Saying Barolo is my "favorite" is very much a misrepresentation of my everyday drinking habits. How often do I drink it? Outside of professional tastings, when I'm buying wine to serve at home or when I order it in restaurants, I probably drink Barolo two or three times a year. Maybe four if I'm particularly flush. That's because the price of a decent Barolo starts at around $60 a bottle, and quickly climbs to well over $100 at a wine shop. Double or triple that price on a restaurant wine list. No, even though I love Barolo, it will always be a special occasion wine.

After Asta del Barolo, I was thinking deeply about greatness in wines. So I decided to make a quick side trip to visit my old exchange family in Pieve San Giacomo. On a whim, I'd asked Daniela, Paolo's daughter, to do a little research to see where her father used to buy his fizzy red wine, and with some effort we located the winemaker. To my surprise, the winemaker was not based in Modena, but rather a couple hours in the other direction, in the Colli Piacentini—the Piacenza hills—a region I'd never heard of.

After getting lost, and refereeing an argument between Daniela and Anna, who was feeling carsick in the backseat, we were finally welcomed into the garage of the winemaker, 80-year-old Antonio, and his daughter, who was in her 40s. Anna became emotional—the last time she'd visited the winemaker was in the early 1990s with Paolo. "I remember you had a goat and it used to like eating the grapes!" she said. The goat, of course, was long dead.

From stainless steel tanks, we tasted his crisp riesling and an enigmatic, dark yellow wine made from a local and arcane grape, ortrugo. Antonio told me that most of his customers come to buy his wine in demijohns because they prefer to bottle it themselves, as Paolo did.

"What about the frizzante red?" I asked. "Do you still make it?"

He smiled broadly and fished a bottle from a corner of the garage. He grabbed a wide white bowl and as he splashed in the deep purple wine, pink foam bubbled up. "My customers insist on white bowls for the red," Antonio said, "to bring out the color and aromas."

I closed my eyes and took a sniff, and then took a sip. Sharp, fresh, tangy, earthy. Wow. The wine in the bowl was a time machine. I was again 19, dressed in Birkenstocks and a Phish T-shirt, experiencing the aromas and flavors of this wine for the first time. Holding this wide bowl to my face nearly brought me to tears in the dark garage. "Ah, lambrusco," I said with a satisfied smile.

Antonio laughed at me. "Lambrusco? No, no, no. This is Gutturnio!"

"Gutturnio?" I said. What the hell is Gutturnio, I thought. I must have said something wrong. Maybe I was having trouble understanding the dialect. "Is that the local name for lambrusco?" I asked.

He laughed again. "No! It's Gutturnio. It's a blend of barbera and bonarda." (And bonarda, for maximum confusion, is the local name for a grape called croatina.)

Wait . . . what? For more than 20 years, I'd been telling myself that my seminal wine experience was over lambrusco. Now I find out that it's a wine called Gutturnio? And how had I never even heard of this wine? It's not like it's new. I later learned that the Romans drank it from round jugs called *gutturnia*, from which the name is taken. Julius Caesar's father-in-law was famous for producing wine from this region.

We sat at a table and ate cheese and meats with the wine, and Anna and Antonio reminisced about the old days. Antonio said he now sold about 4,000 bottles per year, about half what he did about 20 years ago. "Ah," he said, "a lot of my customers, they're dying." Meanwhile, the younger generation just wasn't as interested in local wines like his anymore. "Nowadays, people want different tastes. Maybe they want cocktails and beer. There are a lot of other tastes that people seek." Antonio shrugged. "There is an end for everything. Everything ends."

Suddenly, this humble purple fizzy Gutturnio that I swirled around in a white bowl—which connected me to my own past, to ancient Rome, and yet at the same time was totally fresh knowledge—seemed more important than even the greatest Barolo. This strange experience I was having in this farmhouse in the Piacenza hills seemed to me to be the very essence of wine, the true reason people spend their lives obsessed with it, an example of how wine becomes part of our lives.

Wine is not a ladder to climb, as we're so often taught. Not even close. Wine is a maze, a labyrinth, one we gladly enter, embracing the fact that we don't know where it will take us, and that we'll likely never find our way out. As I trekked deeper into the maze, I veered away from the so-called Serious Wines, moving further off the beaten path. A larger, more exciting world of wine opened up to me. I began spending more time with grapes like godello from Galicia or teroldego from the Italian Dolomites or vranec from Macedonia or trousseau from Jura or schioppettino from Friuli-Venezia Giulia. Or many, many others.

In my wine articles, I began recommending unfamiliar wines like the sensual reds made from Greece's indigenous grapes, such as xinomavro from Naoussa, in the northern region of Macedonia. Yes, perhaps it seems a bit harder to pronounce xinomavro [ksee-NOH-mah-vroh] than to say merlot or sauvignon blanc or tempranillo. But when you really think about it, it's not. Xinomavro—which literally means "acid black"—is a diva grape, finicky to grow, but when it's good, people liken it to nebbiolo, just like people were bidding on in the Barolo castle. That's because xinomavro is earthy and complex, light in color, with strong tannins and good acidity, and aromas of berry and rose and even tar, similar to Barolo. You'd actually be hard-pressed to discern a young nebbiolo from a xinomavro wine from Naoussa—and if you did, it's because you sensed that the Naoussa wine is a little kinkier than its staid Piedmont cousin.

But here's the twist: Xinomavro will cost you around $15 to $20, about a third to a quarter of a Barolo's price. You could open the xinomavro on a Tuesday night, over tacos made from leftover pork roast, and not feel bad about finishing the bottle while binge-watching *Broad City*. That, after all, is part of the point. Wines made with off-the-beaten-path grapes are fascinating and eccentric, but they are also practical: They repay an adventurous drinker by providing excellent value.

When I think of interesting-but-affordable wines, I think of southwest France, and places like Dordogne, Garonne, and Gascony. These regions are only a short drive from Bordeaux, but instead of cabernet sauvignon and merlot and sauvignon blanc, the best wines here are made from négrette, tannat, mauzac, fer servadou, and petit manseng. No, I am not just making up gibberish words, I promise. Those are names of grapes that come from Fronton, Madiran, Marcillac, and Gaillac. Those are the real appellations where these wines come from. Even though southwest France is that nation's fourth largest area in terms of wine volume, we see very little in the United States. That's not because Dordogne or Garonne or Gascony are upstart, up-and-coming, *nouveau* regions. They're actually quite ancient. Winemaking, in fact, flourished here with the Romans, long before it did in Bordeaux.

For centuries, people have been making big, jovial, rustic reds around the Gascon village of Madiran, where the tannat grape is king. Tannat had a small blip of notoriety in the mid-2000s when scientists found that it contained the highest, most potent levels of polyphenols, those antioxidants that prevent an array of health problems like cancer, heart disease, and diabetes. But it's still largely unknown. Madiran wines are muscular, dark, juicy wines—with notes of black cherry, black olive, and black coffee—that feel just right for when the weather turns cold and leaves begin to fall. Bottles of Madiran washed down a decadent meal in Gascony where I was fed so much rich foie gras and pressed duck that I feared my host might be preparing me for the same fate as the geese. Just about every Madiran I've seen retails for under $25, most under $20.

At the other end of the spectrum are the lighter-bodied, blood-purple wines made with the fer servadou grape, which is called braucol in Gaillac and mansois in Marcillac, its two main growing areas. Fer servadou (or braucol or mansois) shares certain food-friendly characteristics with my beloved Loire Valley cabernet franc, but the savory elements, such as eucalyptus and dried rosemary, are just a little more intense, with a fresh finish like the best shower you've ever had, kind of like those old Irish Spring commercials where the suave guy in the turtleneck sweater wins the Celtic lass. Every Marcillac I've ever opened is always drained very quickly: the most reliable test of a wine's drinkability. Most Gaillac or Marcillac wines retail for under $25.

Perhaps my favorite wines from southwest France are those from Fronton, made with négrette, a delicate red that's perfumed like a Mediterranean garden with exotic dried herbs and wild plums. It feels like a wine that might have been served in Byzantium—which makes sense since négrette is related to xinomavro, and according to legend it was brought back to France from Cyprus during the Crusades. Just about every Fronton I've seen is priced under $20.

I could go on naming more grapes, slipping further down the maze. But at this point, you might ask: Why can't I find affordable wines like these on more store shelves and wine lists? Why aren't more people talking about interesting, good-value wines like these? Why aren't grapes like fer servadou and tannat and négrette found in more regions of the world, the same way the noble grapes are?

Perhaps it's because tannat is a very tannic grape, which makes it more difficult to work with than, say, merlot. Maybe farmers stopped planting fer servadou because the vine's hard woodstock (*fer* means iron in French) was less convenient than cabernet sauvignon. Perhaps négrette was too finicky and delicate. Or, perhaps, grapes like fer servadou and tannat and négrette are obscure for a much simpler reason: Powerful people have always wanted it that way.

Back in the 13th and 14th centuries, the merchants in Bordeaux began to see the wines of southwest France as a threat to their economic interests. So Bordeaux, wanting to keep its dominance over the wine trade with England, flexed its muscles to establish a strict code called *police des vins*, which decreed that no wine could be traded out of Bordeaux until the majority of Bordelais wine had already been sold. This dealt a devastating blow to the winemakers in southwest France. In Hugh Johnson's book *Vintage: The Story of Wine*, he quotes Anne Robert Jacques Turgot, finance minister under Louis XVI, who described the effects of *police des vins* "The conduct of this set of rules, most artfully devised to guarantee to the bourgeois of Bordeaux, the owners of the local vineyards, the highest price for their own wines, and to the disadvantage the growers of all the other southern provinces."

When the winemakers in southwest France could no longer sell wine at premium prices, those areas settled into a provincial backwater. Some of the growers ripped out their vineyards of local grapes and replanted the noble grapes, in an attempt to ride Bordeaux's coattails. Those that continued to work with local grapes made low-cost, everyday "peasant" wine. The rest of the world mostly forgot about fer servadou and tannat and négrette. Due to Bordeaux's power and influence, it's taken more than 500 years for wine geeks to rediscover the indigenous wines of southwest France.

Even today, there are plenty of powerful people in the wine world who fight to maintain the status quo of the noble grapes, and they throw shade on the idea of exploring new grapes or rediscovering classic-but-forgotten regions. One of them is Robert Parker, the so-called "Emperor of Wine," who established his sizable reputation and fortune by recommending the wines of Bordeaux and Napa Valley to Baby Boomers who came to wine in the 1980s and 1990s. Not too long ago, he may have been the world's most influential critic on any topic. Now in his 70s, his influence among younger wine enthusiasts on the wane, he is still a force to be reckoned with—feisty, combative, and antagonistic toward any point of view that runs counter to his.

The clearest and most notorious example of Parker's prejudice toward non-noble grapes happened in early 2014, just as I'd begun sharing my discoveries of grapes such as xinomavro or fer servadou in my earliest wine articles. Parker published a now-infamous, unhinged rant that scolded people like me who embrace off-the-beaten-path wines. He called us "Euro-elitists" and "wannabes" and "extremists" whose recommendations are "the epitome of cyber-group goose-stepping" and "Kim-Jung-unism." Yes, seriously. He did. It was the literary equivalent of an old man shaking his fist at younger punks.

The rant's headline—"There Is No Reason And The Truth Is Plain To See"—was taken from the lyrics of Procol Harum's 1967 hit "A Whiter Shade of Pale." Four times in the first 700 words, Parker referred to his "35 years" in the wine-critic business. He spewed out a lot of lines that sounded as if they were taken from 1950s anti-communist propaganda, full of "perpetrators" and "neo-intellectuals" and suggested that those who offer points of view contrary to his are like the "propaganda machines of totalitarian regimes." He also said our way of writing about wine needs to be "condemned" and "repudiated."

Mainly, what Parker seemed riled up about is the growing number of younger wine writers and sommeliers who don't share his opinions. Some of us like so-called natural wines;

some of us are fed up with high-alcohol, over-oaked, jammy fruit bombs; some—like me—just happen to recommend lesser-known wines to our readers. Most of us have grown bored of the expensive, prestige Serious Wines. What struck me most were these passages:

> *What we also have from this group of absolutists is a near-complete rejection of some of the finest grapes and the wines they produce. Instead they espouse, with enormous gusto and noise, grapes and wines that are virtually unknown. That's their number one criteria—not how good it is, but how obscure it is . . .*

> *Of course, they would have you believe some godforsaken grapes that, in hundreds of years of viticulture, wine consumption, etc. have never gotten traction because they are rarely of interest (such as Trousseau, Savagnin, Grand Noir, Negrette [sic], Lignan Blanc, Peloursin, Auban, Calet, Fongoneu, and Blaufrankisch [sic]) can produce wines (in truth, rarely palatable unless lost in a larger blend) that consumers should be beating a path to buy and drink.*

I was sort of impressed by Parker's level of bile. I don't think I've experienced a rant like that one since I was 15 and our neighbor in the wifebeater called me and my skateboarding friends "dirty longhairs" and shouted at us to get out of his driveway. But there it was, in black and white. Power had spoken. The biggest gatekeeper in wine had excoriated people like me for our perceived "near-complete rejection" of grapes like chardonnay and cabernet sauvignon or pinot noir. He scolded us for recommending what he called "godforsaken grapes." And he even named, among the godforsaken grapes, several that I had recently written about, including négrette, which Bordeaux had tried to eradicate hundreds of years before.

Of course, my poor son Wes, as he put the finishing touches on his safety speech at the breakfast table, had no idea that his questions tapped into all of this. Nor did he know that his father had been cast as some sort of wine outlaw—a lover of "godforsaken grapes"—by the world's most influential wine critic.

What Wes knew that morning—and what was upsetting him more than spelling errors—was that after I walked him to school, I would be going away. Again. That afternoon I'd be flying off to Europe, to taste and discover some more godforsaken grapes—in some cases the very grapes that Parker had singled out for scorn. By that time the next day, I'd be back in Switzerland, tasting savagnin. Not long after that, I would be in Austria, tasting blaufränkisch. And only a little while later, I would be in southwest France, tasting négrette. Among many other, even more uncommon grapes. Why was I doing this? At that moment, I couldn't have possibly explained. I was already obsessively lost in the wine labyrinth. But I couldn't tell you why this obsession with strange, unpronounceable grapes had caused me to leave my secure job, destabilize my family, and pursue this mania. All I could say was that it felt important.

"Dad, when are you coming back?" asked Wes, as he slid on his backpack and popped more concord grapes into his mouth.

"A few weeks," I said. He hugged me tight and buried his head into my chest so I didn't see his tears. "It'll be OK. I'll be home before you miss me." When he finally released me, I had the tiniest purple grape stain on my shirt.

CHAPTER THREE

Wine and Dada

SWITZERLAND IS WELL KNOWN FOR MANY THINGS: Swiss chocolate, Swiss watches, Swiss cheese, secret Swiss bank accounts. But Swiss wine? Most people don't even realize there is such a thing as Swiss wine. The Swiss certainly don't make a whole lot of wine, only about a million hectoliters—a drop in the bucket compared to France's 42 million or Italy's 48 million hectoliters. And Switzerland only exports about 2 percent of its wine. By comparison, Italy and Spain each export about half of the wines they produce. So according to just about any metric, Swiss wines represent the height of obscurity. Yet nearly 40 indigenous varieties grow in Switzerland. So what better place for my obsession with enigmatic grapes to take me.

I immediately fell for Switzerland's most-planted grape, chasselas. Chasselas, some say, is the perfect wine for ten o'clock in the morning. Others say it's the perfect wine to drink when you are really thirsty. I found both of these notions to be true on my second morning in Zürich, ordering a glass after my croissant and coffee at a sunny café table in the medieval Old Town. I can report that chasselas is the perfect wine to follow a double espresso, and goes down way too easily. By 10:15, I'd ordered a second glass.

The Swiss love their chasselas, even though the rest of the world mainly just eats the grape or turns it into juice rather than wine. Chasselas wines are often described by non-Swiss as bland, soft, or neutral, and they're often demonstrably low in acidity. "I always get a shock, when taking my first mouthful, at how low in acidity it seems," once wrote esteemed British wine critic Jancis Robinson. The best chasselas offers the drinker a unique drinking experience—dry, delicate, a little chalky, a tiny bit salty, a smidgen floral, and sometimes even milky or smoky. Swiss wine critic Chandra Kurt describes the acidity of high-quality chasselas as "just the right amount—and by right, I mean inconspicuous," adding that it's a wine "of modesty and understatement," like Switzerland itself.

"It's hard to convey just how important chasselas is to Swiss culture," José Vouillamoz had told me the year before, at the raclette dinner in Sierre. "You have chasselas at a wedding, at a funeral, to close a business deal, to make a political compromise. At all occasions you have chasselas." He added, with a wink and a laugh: "Except for food."

My second glass of chasselas was from Valais, where I'd been with the Wine Mosaic crew a year earlier. In Valais, the grape is called fendant, and grows especially fine near the town of Sion. A century ago, Fendant de Sion was one of Europe's sought-after wines. James Joyce, who loved Fendant de Sion when he lived in Zürich, affectionately (and a tad pervertedly) called it "the archduchess's urine."

Ernest Hemingway, always particular and exacting about drinks in his work, wrote several stories set in the Alps in which his characters call for "a bottle of Sion." In *A Moveable Feast*, he writes about drinking Sion wines amid wistful, regretful memories of his first wife, Hadley. In "A Cross-Country Snow," young Nick Adams (Hemingway's autobiographical stand-in) orders "a bottle of Sion" with his friend George near the end of a skiing trip. After the German-Swiss waitress has trouble with the cork ("those specks of cork in it don't matter," says Nick) the Sion wine serves as the catalyst for the mood shift from boyish banter to something gloomier, the passing from youth into the world of adult responsibility.

> *"Wine always makes me feel this way," he said.*
> *"Feel bad?" Nick asked.*
> *"No. I feel good, but funny."*

I was feeling similarly funny. Maybe it was the second morning glass of chasselas. Or maybe it was the skinny, pungent cigarillo I'd lit and was now puffing, the one I'd bought from an old-fashioned tobacconist, Tabak-Lädeli, across the river from the Rathaus, near Saint Peter's 14th-century clock tower. Or maybe it was because I was rereading—puzzling my way through, really—Rainer Maria Rilke's peculiar, often-impenetrable *The Notebooks of Malte Laurids Brigge* on my iPad. Or maybe it was simply Zürich. I hadn't been in this neighborhood for more than 25 years, since I was 19, traveling with a friend from Vermont named Sara, a copy of Rilke's *Letters to a Young Poet* tucked in my backpack. I'm not normally a smoker, so perhaps I'd bought the cigarillos at Tabak-Lädeli today because that's what I'd done way back in 1990. A semi-famous poet had visited my undergraduate creative writing class—unshaven, corduroy jacket with arm patches, beret, and smoking a cigarillo. At that callow age, I was impressed, and endeavored to live that kind of louche poet's life. I remember Sara and me leaning out a second-floor window of a shabby hotel overlooking Niederdorfstrasse, smoking, drinking cheap Swiss wine, skimming Rilke, and pretending to be cosmopolitan.

Funny, I guess, to associate all this with Zürich, which many people consider to be so quiet and clean and staid and bourgeois, a watch ticking inaudibly on a banker's wrist. In an essay about his beloved Zürich, pop philosopher Alain de Botton writes that the city's "distinctive lesson to the world lies in its ability to remind us of how truly imaginative and humane it can be to ask of a city that it be nothing other than boring and bourgeois." De Botton insists that what's most exotic about

Zürich is how "gloriously boring" it is. "We normally associate the word exotic with camels and pyramids. But perhaps anything different and desirable deserves the word."

In any case, I want to be clear that chasselas is the perfect wine to drink along with tobacco smoke, on a sunny morning in Zürich. The Swiss guy at the café table next to me, who was smoking a pipe and also enjoying a glass of chasselas, seemed to agree. I was too distracted to make it very far into *The Notebooks of Malte Laurids Brigge*. "I am learning to see," Rilke writes. "I don't know why it is, but everything enters me more deeply and doesn't stop where it once used to." I felt like I was finally learning to taste, and something similar was happening within me. My palate had guided me to Switzerland, and ever deeper into the wine labyrinth.

In reality, the most likely reason I felt funny was due to the amount of fermented grape juice I'd consumed the night before. I'd done a tasting of Swiss wines with a large, gregarious man named Philipp Schwander, an importer and one of Switzerland's only certified Masters of Wine. I'd met Schwander months before through my friend Stefano, an Italian winemaker in the Colli Berici in Veneto. Schwander imports Stefano's wines into Switzerland, and he arrived at the 17th-century villa squeezed into a tiny, expensive convertible sports car.

During a tour of the villa, Schwander said, "Stefano, I am now the proud owner of a castle, as well. In Lindau, on the Bodensee." We all were supposed to taste Stefano's wines over lunch at a trattoria, inside a hotel in the Berici hills. Schwander, however, looked pale and sweaty and confessed that he wasn't feeling so well. He claimed it must have been dinner the night before at a Michelin-starred restaurant near Verona. "I believe it was the sous vide meat," he said, referring to the trendy technique of cooking food, vacuum-sealed in plastic, slowly in a warm water bath. "I never quite trust this new way of cooking. Cooking meat in a sealed bag at lukewarm temperatures!" He insisted that drinking a Campari before lunch would cure him. When that didn't work, he asked for a room in the hotel to take a nap.

Now, six months later, at his office in downtown Zürich, Schwander was in much better health and spirits. He greeted me warmly and poured a half-dozen Swiss whites and a half-dozen Swiss reds. These were incredibly scarce wines. In fact, the entire annual supply of Swiss wine can satisfy only 40 percent of Switzerland's wine demand, and so the rest of the world is lucky to taste any of it. So I was very excited and grateful for Schwander's hospitality.

He began with a zingy, complex müller-thurgau from around Lake Zürich. "They made this special wine for us, with a special yeast that they found in a bottle dating from 1895," Schwander said. Next, we moved on to a couple of chasselas. "It's fashionable to say that chasselas is drinkable, but that it cannot make a wine of high quality," he said. "But that's not true." My favorite was a fendant from Valais, from the grand cru surrounding the mountain town of Vétroz. It was muscular, rich, and creamy—an Alpine slalom-champion sort of chasselas.

The next wine Schwander poured was made from a grape called heida—fleshy, curvy, a little spicy, and a little nutty. Heida is the local Valais name for a grape that people in the German-speaking world call traminer. For years, it was believed the grape originated in the Tyrolean village of Tramin, where it took the name traminer. But that has been recently called into question. The grape is also notably grown in the French region of Jura, near the Swiss border, where it's called savagnin (not to be confused with popular *sauvignon*; yes, the parsing of grape names can induce a migraine). Jura savagnin has become a darling of many younger sommeliers, popping up on new-wave wine lists, thus earning the scorn of Robert Parker, who trashed savagnin among his "godforsaken" varieties.

Heida—or savagnin or traminer—is genetically unstable, mutating easily. Spontaneous mutations can happen with any growing plant as its cells divide. But with grapes, sometimes these mutations (higher yields, different coloring, larger or smaller berries) appeal to winegrowers, and they propagate the wild mutations into clones. Numerous clones of heida/savagnin/traminer have been propagated throughout the Alps since at least the tenth century, and some clones are thought to be the genetic parents of sauvignon blanc, chenin blanc, and grüner veltliner. No one, though, truly knows the grape's origin.

As it happens, the chasselas and the heida that Schwander offered me were made by Domaine Jean-René Germanier, a producer I'd visited the year before on my trip through the Alps with Wine Mosaic. Inside Domaine Jean-René Germanier's modern-design tasting room, a few miles from Sion, I met the winemaker Gilles Besse, who'd joined his uncle in the family cellar after 15 years as a jazz saxophonist and band leader. Besse was in his 40s like me, but much taller, fitter, and more tanned, wearing a vest and sunglasses, and seeming like the essence of The Alpine Man. He drove us in his luxury SUV up into his mountain vineyards. At 2,500 feet above sea level, we got out and wandered the Clos de la Couta vineyard. The sun was amazing. Valais is the sunniest place in Switzerland, with more than 2,500 hours of sunlight every year, more than Burgundy, Bordeaux, and many other European regions. The light was so brilliant, it threw everything into stark relief. We could have sunbathed between the vines.

Valais is a place of small producers—altogether, about 22,000 growers work 12,000 acres, most working less than a half acre, some only a few rows. The smallest, in the village of Saillon, has only three vines and allegedly belongs to the Dalai

Lama. "Every family in Valais has vineyards, so even when it was hard to sell the grapes, it was a side job. They weren't professionals. And so people just kept the old grapes." Even today, Domaine Jean-René Germanier still grows 14 different varieties, including rarities such as petite arvine, humagne rouge, and amigne. "In the mid-1980s, my generation said, 'We have to plant local varieties.' But it wasn't with a nostalgic spirit. We still grow pinot noir and syrah. We just asked the question: What is our history, and what can we take forward?"

Still, I remember most deeply the obscure Swiss varieties that Besse poured us. The humagne rouge, with its rustic elegance, tasted honest and savory with attractive peppery and smoky notes—sort of like pinot noir if it took up smoking a pipe. Amigne, of which only about 100 acres exist, is so ancient that some believe it was brought by the Romans. And if the Romans didn't bring it, amigne probably was already here growing wild. "We don't know its exact parents," Besse said. "It has small berries and long grapes. But it's without seeds and without flowers. It has . . . how do you say . . . a sexual problem," he said, with a laugh. (Yes, as you are learning, wine people love double entendres and sexual innuendo.)

In any case, amigne, with its sexual problems, is the type of weird, fussy grape that a profit-focused, corporate winery selling mainstream wines would rip out in a heartbeat.

Finally, we tasted Besse's heida, which is grown around the village of Vex and so is labeled Heida de Vex. Perhaps unsurprisingly, as we enjoyed drinking the heida (aka traminer, aka savagnin) the name Robert Parker popped up. Besse was confounded by the rant, since Parker's *Wine Advocate* has given his syrah wines very high scores. Over a year later, Besse was still irritated at the shade thrown on his beloved grape savagnin. "He goes completely against the trend," Besse said. "He likes to portray *obscure* as uninteresing. But I think it's simply to protect his commercial interests."

Standing now in Schwander's Zürich office, the Heida de Vex I swirled in my glass evoked brilliant sunshine in the mountains, like one of those rare, sunny spring days I remembered from college in Vermont, where there was still snow on the ground, but everyone skied in T-shirts. I didn't want to quite move on from the heida, but two friends of Schwander's arrived, one an insurance executive and the other an energy company CEO. They all suggested we move on to the reds. Schwander opened several lovely Swiss pinot noir, as well as two merlot from the canton of Ticino, the Italian-speaking part of Switzerland. All of the reds were good, especially the strange, austere pinot noir from near Lake Zürich. But I'd had so many other nice, or *nice enough*, pinot noir in my life. My mind and heart were still fixated on the lesser-known grapes.

Christian, the insurance executive, told me he owned a winery in Valais outside of Sierre, where I'd gorged on raclette. We expressed our mutual admiration for the "Temple of Raclette" and Christian told me his winery was not too far from the Foundation Rilke. Rilke, he said, had lived out his postwar years in Sierre, from 1919 until his death in 1926, in a château owned by one of his wealthy patrons. Now his papers are preserved there, surrounded by the vineyards of Valais. In Sierre, in a "savage creative storm," Rilke finished his masterpieces *Duino Elegies* and *Sonnets to Orpheus*.

"Rilke!" I said. "I've actually been reading him this week!"

"Do you read Rilke in English?" asked Christian. "It's so difficult to translate lyrical poetry."

"It's sort of like wine, no?" Schwander said. "Very difficult to translate as well."

Rilke might have appreciated the scenario of this conversation. He was a lifelong moocher and dandy who enjoyed the good life, paid for by wealthy friends and patrons. And the "unsayable" was an idea he'd certainly covered long ago. In my old tattered copy of his *Letters to a Young Poet*, I'd underlined, in my youthful hand, this passage: "Things aren't all so tangible and sayable as people would usually have us believe; most experiences are unsayable, they happen in a space no word has ever entered." Tasting uncommon wines and explaining them would seem to fall into this category.

As the office tasting wrapped up, we climbed into Schwander's big black Mercedes-Maybach sedan. The back seat was so huge that I could almost fully recline. There were soft pillows on the headrests and TVs on the back of the driver and passenger seats. We drove out of Zürich, down the lake, to the hilly suburb of Gattikon, to Smolinsky's Sihlhalde, a Michelin-starred restaurant. I asked Schwander if he was still wary of sous vide cooking. "Well, I like this place because they still do traditional cooking. No sous vide here." We dined on veal cheeks and truffle-stuffed ravioli and Schwander opened some really expensive bottles of Burgundy and vintage Champagne. There was some talk about the comparative differences between 50-euro and 250-euro Champagne and their merits. Later, when I told my friend Stefano about my evening in Zürich with Schwander, he shook his head and said, "The Swiss! They have so much money. But they are so bored. Hey, let's eat at expensive restaurants and drink expensive wine and drive around in expensive cars!"

After dinner, right at the table, Schwander unrolled architectural blueprints for the renovation work that was happening on his castle at Lindau, on the Bodensee, otherwise known as Lake Constance. I had been to Lindau once, years ago, on my college backpacking tour with Sara. We did not stay in a castle, but rather a smelly hostel. In fact, Lindau is where, after a week of sharing a bed in some kind of cool, complicated hippie-platonic Alpine idyll, the relationship came to a drunken head, with feelings expressed, feelings hurt.

Back at the sunny morning Zürich café, I finally put down the Rilke. I finished the morning chasselas and cigarillo, took another espresso, paid my bill, then wandered down the narrow cobblestone streets of Old Town. I poked my head into Cabaret Voltaire, which a century before had been the spiritual home of Dada, where the original Dada manifesto was delivered. This year coincided with Dada's 100th anniversary, which Zürich quietly celebrated. The Cabaret Voltaire is now a café, bookstore, and art space, and a small video exhibition in the basement explained that the Dada artists and poets rejected the certainty of Enlightenment thinking, questioned logic and reason, and wanted to "bring irrationality and magic and chance into life." Dada art was silly and childish, as it was meant to be. After all, the name *dada* was supposedly synonymous with a baby's first words, or maybe was the French word for *hobbyhorse*, or maybe was chosen at random. No one knows for sure. At the Cabaret Voltaire, Dadaists shouted nonsensical poems, wore ridiculous costumes, danced crazily to discordant accordion music, and presented discarded garbage as "readymade" art. After a few years, the Dada movement imploded and disappeared. Dada, we are told in Intro to Art History, was important. As one learns, it was a bridge from modernist movements like cubism and futurism that led to surrealism and abstract expressionism and pop art and all the other postmodern movements. But Dada itself was incomprehensible to nearly everyone, including other Dadaists. Its demise was inevitable. What Dada did to outrage people one week was old hat by the next, and so a fresh outrage was always necessary: *OK, so last Tuesday we told everyone that urinal was "found" art . . . what the hell can we do to top that today?* I occasionally worry that the pursuit of ever more obscure and lesser-known wines is sort of like Dada. What's cool and enigmatic one day—trollinger from Germany or encruzado from Portugal or malagousia from Greece—could very well become boring tomorrow. For all I know, Fendant de Sion, "the archduchess's urine," will be poured by the glass in every hipster wine bar by the time this book is published.

And yet, you often hear wine people talk high-mindedly about wine as approaching "art" and winemakers as akin to "artists." If this is the case—and I'm not convinced it is—wine-as-art would still basically be stuck in the 19th century. Wine as a cultural discipline hasn't even left the Enlightenment, with wine critics and certified sommeliers and other gatekeepers still playing the role of the old Academy, with their judgmental faux-certainty and numerical scores and false logic that ranks and categorizes which wines are deemed "best." In some cases, wine culture hasn't yet even encountered modernism, let alone postmodernism. Perhaps wine needs Dada.

Oy, with these kinds of thoughts rattling around my head, I realized I needed another glass of wine, quickly. It was lunchtime, so I wandered up Rindermarkt, to Oepfelchammer, a classic wine tavern that's been around since 1801. Inside, dark wood beams, paneled walls, and long wooden tables with benches have hundreds of names carved into them over two centuries. The waiter, Boris, sported a pointy mustache and beard, looking like Trotsky. He brought me a surprising house white, made from räuschling. "This is a very old variety that grows around Lake Zürich," Boris said.

In fact, räuschling was first grown in the Middle Ages in the Rhine valley of Germany, believed to be the love child of gwäss (or gouais blanc) and savagnin (or heida or traminer) and was first mentioned by the important 16th-century German botanist Hieronymus Bock. Less than 60 acres of räuschling exist in the world. The name is believed to have come from *rauschen*, the sound of the wind rustling the dense leaves that characterize this grapevine. Being named after the sound of wind made sense: Drinking räuschling was more about a feeling than particular flavors. This was an odd wine, with super-crisp acidity, little discernible aroma, and a dose of saltiness. I wondered what would pair with this wine.

I took a seat by the open window, which overlooked Rindermarkt, next to yet another Swiss guy smoking a pipe. Three sporty young men—German-speaking bros—drank and ate boisterously at a nearby table. I ordered chanterelle soup and classic Zürcher Geschnetzeltes, sliced veal strips in a mushroom cream sauce, served with a side of rösti, the famed German-Swiss take on hash browns.

Boris suggested two different reds, one called cornalin and another called cornalin du Valais. Reaching the zenith of wine-naming insanity, Boris explained that these two cornalin were actually completely different grapes. Cornalin du Valais, he said, is actually local Valais slang for a grape called *rouge du pays*, or "local red." Boris said the other cornalin—which he called simply *cornalin*—was another name for humagne rouge, the rustic wine from Valais that tasted like the hick cousin of pinot noir. In fact, most cornalin is grown across the Alps in the French-inflected region of Italy called Vallée d'Aoste. I chose the cornalin du Valais, or rouge du pays, which was dark purple, peppery, and smelled a bit like incense.

Just as I was tucking into my Zürcher Geschnetzeltes, one of the sporty young men rose from the boisterous table. Encouraged by his buddies, he jumped and grabbed the huge ceiling beam and tried to pull himself up. On the menu, I'd seen an illustration of what was happening: This was Oepfelchammer's traditional *Balkenprobe*, a quirky challenge to "take the beam." Challengers must climb up over one beam headfirst, crawl over a second, and then hang by their legs, where they are then given a glass of wine to drink upside down.

As the young man struggled to pull himself up and through the beam, everyone in the tavern—two couples, the guy with the pipe, and me—knocked their fists on tables in encouragement. Boris stood below him, as a spotter. "It's all about the technique," he said to me, in English. Finally, red-faced, the young man pulled himself across the second beam, hung down, and Boris poured him a glass of the räuschling, which he finished in a few gulps. Then he swung to the ground, and returned to his cheering mates at the table.

Watching this spectacle did answer at least one of the questions rattling around inside my head that day: Räuschling clearly pairs well with hanging upside down from a 200-year-old wooden beam. This, of course, still left plenty of other grape mysteries to puzzle over.

CHAPTER FOUR

Alpine Wines

WHEN I WAS A KID, I had a very good memory for sports trivia and statistics, particularly baseball and basketball. I was the guy to go to if someone in my family wanted to know, say, how many RBIs Mike Schmidt had during his MVP year for the world champion 1980 Phillies (121) or how many times Moses Malone led the NBA in rebounding (six) or how many players the Phillies traded for Von Hayes in 1982 (five: Julio Franco, George Vukovich, Jay Baller, Jerry Willard, and Manny Trillo)—which, to be honest, was almost never. But I can remember spending hours geeking out over *The Bill James Baseball Abstract*, and other early books on esoteric, advanced baseball statistics, now known as sabermetrics (made famous by the book and movie *Moneyball*).

As I went deeper and deeper into the realm of obscure and esoteric grapes, keeping track of all this arcane nomenclature and information, I began to see parallels with my old love of sports trivia. It often seemed to satisfy the same parts of my brain. I also began to see ampelographers as new heroes, potentially changing how wine is experienced and enjoyed in the same way that sabermetrics changed baseball. I consulted the 1,242-page *Wine Grapes* by José Vouillamoz, Julia Harding, and Jancis Robinson in the same way I used to pore over Bill James. *Wine Grapes*, with its colorful early–20th century illustrations of grapevines, may have become even more important, transcending into a sort of holy book.

I thought often of Jean-Luc Etievent and Petit Verdot (Jean Rosen) and my trip with them through the Alps. These guys were total wine geeks, and they'd reached a level of geekdom I could scarcely imagine. I remember standing with Etievent and Petit Verdot inside of an 11th-century abbey, in the medieval town Saint-Chef, in the Isère region in southeast France, near Grenoble. Joining us was Nicolas Gonin, a highly regarded younger Isère winemaker. Most visitors focus on the abbey's renowned 12th-century Romanesque frescos. But the four of us were looking at the ceiling. Etievent pointed to the frieze depicting bunches of grapes on a vine. "This is your first ampelographic question," he said. "See the grapes and leaves? What is the variety?"

I sure as hell didn't recognize them. Etievent then asked Gonin what they were. "I don't know," he said, with a smirk. "But I've been looking for them for a long time."

Etievent's little quiz in the abbey was a trick question—the vines depicted in the frescoes had likely not been seen in this region for almost a century, perhaps longer. Vineyards have surrounded Saint-Chef for more than a millennium, with the first extant mention of grapes dating to 993. "We don't know if there were already vineyards here before the Romans, or whether the grapes are a cross between local and Roman varieties," Gonin said.

"But no one knows the wines from here anymore," Etievent said. "Phylloxera destroyed everything." By phylloxera, he meant the devastating plague of sap-sucking, almost microscopic insects that feed on the roots of grapevines, killing them. These tiny yellow aphids are a key reason why scores of local grape varieties started to disappear. Beginning in the mid–19th century, phylloxera ravaged vineyards across France and the rest of Europe. From the late 1850s to the mid-1870s, almost half of France's vineyards were destroyed.

As their vines shriveled, wine growers tried hundreds of remedies to halt phylloxera, pumping and spraying gallons of chemicals and pesticides and sulfur treatments onto leaves and into the soil. The French government offered a 300,000-franc reward to anyone who developed a cure for the blight. Desperate growers even tried flooding their vineyards with seawater or allowing their chickens to roam the vine rows in hopes they might eat the insects. Local priests sprinkled holy water onto the soil. One priest in the Loire Valley told farmers to position four special crosses—which the priest conveniently sold—around each vine to ward off the scourge. He sold more than 10,000 crosses before people caught on that his scam didn't work. The most infamous solution involved burying live toads under each vine to draw out the "poison." Nothing worked.

From the beginning, the French claimed that the blight was introduced to Europe from America, and that's now accepted as fact. Phylloxera has always existed in North America, but our foxy native species such as *vitis labrusca* have been resistant to its damage. With the advent of faster transatlantic steamship travel, the pests could now survive the ocean crossing. And so the ugly American scourge arrived to kill off European *vitis vinifera* varieties.

By the 1870s, it was clear there was no cure. The only surefire, preventive solution was to graft European vines onto rootstock from North America that was already resistant to phylloxera infestation. This is done by taking a thick, older trunk of a North American vine and fusing it with a new, young European *vitis vinifera* vine (called the scion). The tissues of the two vines are joined, sealed, and soon begin to grow together.

Though this grafting process was—and still is—incredibly effective in managing phylloxera, it didn't sit well with French winemakers. Many refused to accept the idea that they'd have to mix their vines with what they saw as inferior North American rootstock. And so they persisted in putting things into their vineyards: chemicals or toads or crosses. They shouted down those growers who grafted as "Americanists." To this day, many people believe that the wines made from the self-rooting vines were of better quality, and they mythologize pre-phylloxera wines. Of course, this stubborn, chauvinistic attitude is why phylloxera continued to devastate vineyards throughout the late 19th century. By 1900, almost three-quarters of France's vineyards were lost.

The Isère region was hit especially hard, and was slow to recover. In fact, more than a century beyond the phylloxera pandemic, there is still no official *Appellation d'origine contrôlée*, or AOC, in Isère, whose wine is mostly bottled simply as regional *vin de pays* or "local wine."

"Winemaking sort of disappeared in this region. So what grapes remained were mostly left alone and weren't replanted with anything else," said Etievent, as we puzzled over the abbey's ceiling. Today, that neglect is why Isère is a cradle of rare, indigenous grapes.

When we'd looked at enough frescoes, we drove to Gonin's simple tasting room in a flat yellow building near Saint-Chef. In 2003, Nicolas, then in his late 20s, took over family land, where his grandfather had farmed vegetables and tobacco alongside the vineyards. At first, Nicolas planted the usual suspects, chardonnay and pinot noir. "If I'd been more courageous, I would have put in more local grapes." Over the next 15 years, he got brave, planting traditional local grapes: whites such as altesse and verdesse, reds such as persan and mondeuse. Now his wines appear on the lists of in-the-know, big-city sommeliers.

The most popular of Gonin's wines among the sommelier crowd is made from a grape called persan. In the 18th century, persan was considered one of the best red wines in France, but now there are only about 25 acres in existence. Gonin's persan is very low alcohol, often under 11 percent, aged in steel or plastic tanks rather than oak barrels, and funky. In fact, if you looked up *funky* in the dictionary, you might see an image of this wine. Which is fortunate for Gonin, because funky, low-alcohol, unoaked wines are currently all the rage among the new generation of sommeliers. "In France, the older sommeliers don't know how to sell a wine unless it's from a famous appellation," he said. "I have much better luck with young sommeliers in New York. They don't necessarily need the prestige, or a famous appellation. They're curious."

I asked him how it felt to be selling more of his local Isère wines in New York than at home. "For a long time, this was disturbing me," he said. "I was asking myself too many questions like this a few years ago. Now, I don't ask myself anymore."

We tasted Gonin's 2014 altesse, which was delicate, honeyed, and a little nutty. Altesse (meaning "highness") is also known by the alias rousette in nearby Savoie and, until recently, people believed it had been brought to France from Cyprus by a Savoie duke in the 18th century, a story similar to how négrette arrived in southwest France. This story has been largely debunked, though, and altesse is now believed to be an indigenous Alpine variety.

Next, we moved on to Gonin's even more rare 2014 verdesse—less than a dozen acres of verdesse exist. It tasted like snow from a mountain meadow that had been collected and melted in a hollow melon. Only around a dozen acres of the grape are grown in the world. "This grape grows only in Isère, and we hope to keep it that way," he said.

"So you don't want to see verdesse grown in California?" I asked.

Gonin looked at me with a smirk, like I was crazy. "No," he said. For Gonin, indigenous grapes are inextricably tied to place and culture. California verdesse would be like Maine lobster from Kansas.

While I sat in Gonin's tasting room, I nodded politely at this idea. But I don't totally agree with it. In fact, one of my favorite grapes, durif, comes from Isère not far from Gonin. Durif was discovered in the 1860s, just before phylloxera, by a botanist and grape breeder named François Durif in his experimental vineyard. Syrah is durif's prestigious parent, which mated with another grape called peloursin, a variety that Robert Parker included in his shaming list of godforsaken grapes. Durif, like verdesse, never was planted widely and by the early 20th century, following phylloxera, it had almost entirely disappeared from Isère, if not all of Europe.

However, in the late 19th century, a winemaker near San Jose, California, imported the durif vine, and changed its name to one that many American wine drinkers will recognize: petite sirah. "They decided to give it a new name, including a French word for prestige, and something a bit easier to spell than Syrah," wrote Roy Andries de Groot, in his 1982 book, *The Wines of California, the Pacific Northwest and New York*. Isère's durif thrived under its new identity in sunny California. Today, there are around 6,500 acres planted in California, up from 2,200 acres in 1999. In 2002, petite sirah producers formed an advocacy group called P.S. I Love You to promote the grape.

The exquisite paradox of petite sirah is that the grape is neither *syrah* nor *petite*. Petite sirah is usually a deep, teeth-staining purple, full of big, viscous blueberry, chocolate, and eucalyptus flavors—what one producer (Villa San-Juliette from Paso Robles) calls "blueberry motor oil." You do not want to spill petite sirah wine on your shirt. Mark Oldman, in his book *Oldman's Brave New World of Wine*, called petite sirah "dark and intense as a dominatrix's boot." (Ahem, yes, another sexual simile.)

In any case, among Serious Wine people, it feels vaguely embarrassing to admit you enjoy petite sirah. Once, in a newspaper column, I recommended an inexpensive California petite sirah as one of several widely available "gateway" bottles for millennials just discovering wine beyond the Franzia box. A sommelier friend immediately chided me on social media: "Are you kidding me, dude? Petite sirah?"

I've often wondered why the wine cognoscenti mostly ignores petite sirah. It certainly has elements they love: It's relatively obscure and confounding, it generally has solid tannins and acidity, it ripens at low sugar levels, and it pairs with hard-to-match dishes (mole or curry anyone?). Really good petite sirah drinks like dolcetto from Piedmont or zweigelt from Austria or baga from Portugal—in other words, uniquely. It's not the grape's fault that overzealous winemakers in California often age the wine too aggressively in oak. Or maybe it's the word "petite" that scares away wine dudes who fancy themselves macho? Or is it that weird, Americanized spelling of sirah, without the *y*, that drives away Francophiles? Certainly the taste, totally distinct from American zinfandel and cabernet sauvignon, isn't for everyone.

In California, for over a century, petite sirah has been grown in the same vineyards as zinfandel and always been used for blending. If a particular year's zinfandel is a little flabby and needs some tannic backbone, or is too light and needs more color, or is too ripe and sweet and desperately needs brightness and acidity—you can bet that petite sirah is added. Under US law, a single-varietal bottling only has to be made from 75 percent of the grape listed on the label. So, if you're a fan of California zin, there's a chance that up to 25 percent of your favorite zinfandel may actually be petite sirah.

"Petite sirah is a great partner with zinfandel. They're picked together, and fermented together," John Olney, COO and winemaker at Ridge Vineyards, told me when I visited him at Ridge's Lytton Estate vineyards in Sonoma. Ridge is one of the few wineries that shares the percentage of petite sirah that's been added to its zinfandel on the label. They have also bottled petite sirah as a single variety since 1971. Olney and I tasted through their library of petite sirah going back 15 years, and the depth, liveliness, and length of vintages like the dark and brooding 2003 or the fresh and vibrant 2006 were eye-opening. Olney told me that, in the 1970s and 1980s, even more petite sirah existed in California than now. "It was really widely planted in Napa," he said. "But when cabernet sauvignon came along, it didn't matter what else there was. Cabernet took over everything. No one likes to hear it, but the gold standard was France, Bordeaux."

Of course, if durif ever did make a comeback in France, the rich irony is that the French producers would probably have to label the wines as . . . petite sirah.

After tasting with Gonin, our group went to lunch at a small bistro near Saint-Chef, where Gonin opened his 2012 and 2010 altesse. The chalkboard listed *cuisses de grenouille* as the plat du jour, and Petit Verdot said, "This wine should be very good with frogs." Everyone seemed really excited about the altesse-and-frog-legs pairing, then crestfallen when the waitress informed us that the chef had no more frog legs to serve.

Following lunch, we dove even deeper into obscurity, if that were possible. We visited a protégé of Gonin's, a producer named Sébastien Bénard, who lives on a rustic farm near the French town of Moirans, just south of Saint-Chef. Bénard, 37 years old, perfectly played the role of Young Hipster Farmer, with a scruffy beard and ripped shorts, and bees from his hives swarming his barn. We drove to see his vineyard, framed by snowy mountains and the city of Grenoble in the next valley. It was easy to see the reason Gonin had mentored him: Bénard had discovered perhaps the only vineyard in the world of an ancient grape called servanin. He had been working his own vines of persan near an 86-year-old man who still maintained his plot.

Bénard noticed an odd leaf shape that made him realize what the man was growing. "Nicolas taught me how to identify the old grapes," Bénard said. "I tried my best."

"Last year, the man did all the work by himself, then he died two weeks before the harvest," he said. Now, Bénard tends to the old man's vines, as well as his own.

"Next year," Gonin said, "Sébastien will make the only 100 percent servanin wine in the world." Gonin beamed proudly at his protégé. "This is what we do," he said. "I've visited 300 vineyards like this, looking for lost grapes. I spent four years studying and learning to recognize the grapes by the shape of the leaves."

"On a scale of one to five," I asked. "How rare is this vineyard?"

"Like a six or a seven." I might suggest an even higher number: Servanin isn't even listed in any of the comprehensive grape encyclopedias, including *Wine Grapes*.

They were lucky to save the servanin. Just last year, Wine Mosaic found a few rows of another extremely rare grape nearby, and they implored the farmer to keep them. When they returned the following harvest, they learned that the man had died and his heirs had sold the land to a developer. The grape was lost.

As the sun began to set over the mountains, we drove through rush-hour traffic in Grenoble and east to the town of Bernin, to visit Domaine Finot, run by another young winemaker, Thomas Finot, who began growing local grapes here in 2007.

We strolled Finot's vineyards, which sit at 1,500 feet and offer a view of Mont Blanc, then tasted his wines in the garage-like space that housed his tanks, barrels, and a "tasting room" section that resembled a frat house basement, with a dart board and an old yellow leather couch. But these were wines no fraternity has ever served. In addition to verdesse and persan, there was a red called étraire de la dhuy, of which there are fewer than 40 acres in the world, a fraction of what existed, even as recently as the 1950s. It tasted like eating fresh cherries while smoking a clove cigarette and burning leaves in the backyard.

Then there was a white called jacquère, which was like a bright, tart lemon fetched from a stone cistern. "Oh, jacquère isn't that rare," Etievent said. *Rare*, of course, is relative. With 2,500 acres planted in the world, there's obviously more jacquère than a grape that only grows, say, in one single row in one particular vineyard. But think about jacquère in relation to, say, chardonnay—grown on more than 400,000 acres worldwide—and you'll see how "rare" and "obscure" can be slippery terms. Suffice to say, you're not likely to see a shelf dedicated to jacquère at your local liquor barn anytime soon.

When some neighboring Isère winemakers arrived at Domaine Finot with their wines, our tasting turned into drinking, and a party. Over voices growing louder, Laurent Fondimare, one of the winemakers who joined us, told me that he only began his Domaine des Rutissons label in 2010. He'd had no prior training as a winemaker. "I'm an administrator, but I wanted to do something with my hands." His wife's grandfather was growing verdesse and jacquère and étraire de la dhuy, and he was getting too old to tend the vines. He told Fondimare: "You take these grapes and make the wines you want."

"Twenty years ago, people didn't drink these wines," Fondimare said. "All the old people here don't like these local grapes. When we started, they said, 'Ah these are "bad" grapes.' Now, they say, 'Oh you make a good wine with bad grapes!' This is a victory."

"It's a hard experience, but a very beautiful experience," Fondimare said. "It feels like we're part of a movement."

I awoke the next morning in Montmélian, in the Savoie region. Hangovers be damned, by 11 o'clock we were off to taste more esoteric wines. Montmélian itself is not so esoteric: Though surrounded by majestic, snowcapped Alps, it is the very definition of "a nondescript town." Inside the regional wine museum, we were given a tour of old vineyard and cellar equipment. "Certainly a lot of things are the same in every wine museum, but this is different," said the woman giving us the tour, pointing to leather bags that were used to harvest grapes in the Alps. "And this," she said, holding up a contraption called a *casse-con*, a wooden box that slips over one's shoulders, used to carry soil back up the mountain after it erodes. She made Etievent slip the *casse-con* over his shoulders to model it. "These are part of our mountain winemaking tradition."

Outside of those two items, the tour guide was right, this was similar to every other wine museum I've ever visited: Huge wooden presses, dusty old barrels, antique pruning shears and other rusty equipment, yellowing 19th-century government posters declaring PHYLLOXERA! But Etievent, Petit Verdot, and I paid special attention to a display of antique vine grafting tools, simple devices used to fuse together the tissues of two grapevines. While grafting certainly saved Europe's vineyards from phylloxera, it also complicated things.

Before phylloxera, a grower could just cut a shoot from an existing vine and replant it. After the pandemic, a professional nursery was required to graft local vines onto the North American rootstock. That's still true today. Phylloxera—never eradicated, only managed—continues to be a threat throughout the world. "The development of the nursery was the first technological intervention in winemaking, the first industrial process," Etievent said. "When you have an industrial process, you lose choice." The grapes most likely sold by nurseries and replanted by growers were, unsurprisingly, the noble varieties—chardonnay, sauvignon blanc, pinot noir, cabernet sauvignon, or merlot—chosen because they were easier to cultivate and market. This homogenization continued well into the 21st century. Only recently have consumers begun to yearn for something different.

We finished our museum tour and walked upstairs to enter the Pierre Galet Center for Alpine Ampelography, named for a nonagenarian ampelographer who wrote a comprehensive dictionary of French grapes. Monsieur Galet is considered by many to be the father of modern ampelography, and his life's research resides in the library. Here, surrounded by old books and files, we tasted wines poured from bottles with white labels reading "*Vinification Expérimentale.*" The grape names were hand-scrawled on each label: salagnin (super purple and chalky), serenelle (super spicy and vegetal), and blanc de maurienne (super . . . just . . . well, weird; bitter almonds with hints of bong resin). Serenelle is so obscure that it's not even listed in the current edition of Galet's dictionary of grapes. In fact, only about 50 liters of each of these wines exist in the world. Tasting them was the wine equivalent of spotting a giant ibis, or reeling in a coelacanth, or discovering a lost Bronze Age tribe.

Taran Limousin, one of the center's researchers, said that the negotiation with the grower of the serenelle, in particular, had been tricky. The man had planned to harvest those grapes early in the fall—as he always had in that vineyard. But once the Pierre Galet Center had identified this impossibly rare grape, Limousin and his colleagues implored the man to wait until the grapes had matured. There was serious bargaining. "You cannot just come into his vineyard and say, 'Hello sir, this is for science,'" Etievent said. "No, that doesn't really work."

Instead, they offered to trade him ten kilograms of quality gamay for ten kilograms of the later-harvested serenelle. "It was a good deal for the winegrower," Limousin said. "Gamay is much more valuable."

I knew my father, who fancies himself a fine-wine aficionado, would be awake this early in Florida, and so I texted him a photo of the empty glasses and the handwritten labels. "Tough work," he texted back. "Are the wines any good?" I put my phone away. As fathers so often unwittingly do, he turned up the volume on doubts that had begun to whisper inside my head. All of the wine I had tasted on this journey had certainly been . . . *interesting*. Many of the wines had been very good, some amazing. But not all of them. And each time I tasted a rescued, indigenous grape that I didn't enjoy, I felt guilty, as if my negative judgment might banish this grape forever. As if I were Noah turning some small mammal away from the ark, because it was weird and smelled like bong resin.

At the same time, another thought nagged at me: Was all this just a privileged exercise in geekiness and arcane trivia? I'd started to worry I was falling down the same rabbit hole as those hipper-than-thou wine snobs who sneer at people who order chardonnay. There's a funny consumer study that was conducted recently at Brock University in Canada. Marketing researchers gave three groups of people the same, identical wine—a basic chardonnay—but they called it by different names. One group was given the wine and told it was from a fictional winery called Titakis, which they claimed was relatively easy for an English speaker to say. The second group was given the same wine but was told it came from a tongue-twisting fictional winery called Tselepou. The third group was given the wine with no name at all. Many participants, especially those who had some wine knowledge, actually declared that the hard-to-pronounce wine tasted best, and valued it two dollars more than the easier-to-say wine and three to four dollars more than the no-name wine. Was I becoming this sort of annoying wine jerk, who loved obscurity simply for obscurity's sake? In any case, why get all cranked-up to save crazy, hard-to-pronounce, long-forgotten grapes? Isn't wine confusing enough already?

After tasting these ridiculously rare wines, we drove from Montmélian to the Château de Miolans, a fortress that dates back more than a thousand years and sits 1,800 feet up the mountains, looking out toward the Italian Alps. Château de Miolans was a prison in the 18th century, known as the "Bastille of the Alps," whose most famous prisoner happened to be the Marquis de Sade. Before he escaped the prison, de Sade wrote in a letter: "Either kill me or take me as I am, because I'll be damned if I ever change." Perhaps those who persist in growing godforsaken grapes should take this as their motto.

It was a warm, sunny October day—definitely not the weather I expected in the Alps. Atop the fortress walls, we attended a tasting of wines by Les Pétavins, a group of eight local organic producers. Someone put out a plate of charcuterie and a cheese called Persillé de Tigne, a favorite of Charlemagne in the eighth century, and we tasted lovely examples of such grapes as persan, altesse, verdesse, jacquère, and a wine called Chignin-Bergeron, made of roussanne.

Our host was Michel Grisard, a tall, ruddy-faced winemaker with a huge smile, about my father's age, whom everyone called "the pope of Savoie." Grisard had been traveling and tasting with us, and he'd seemed like a goofy troublemaker, always filling up glass after glass and rushing us on to the next obscure wine. Now it was his turn to share the wines under his label, Domaine Prieuré Saint Christophe—in particular the grape mondeuse noire, which he almost single-handedly saved from extinction in the 1980s. Tasting Grisard's mondeuse noire was a revelation: floral, fruity, smoky, foresty, but so light and dangerously drinkable. In years past, it was called *grosse syrah*, because of the grape's similarity to syrah (of which it is either a half-sibling or grandparent; it's unclear). But mondeuse noire has an edginess, a wildness, that's quite different from syrah. Later at dinner, when we tasted older vintages of Grisard's mondeuse, dating from the 1990s and 1980s, my appreciation only deepened. These were simply great wines. And mondeuse noire just happens to be the most prevalent of three mondeuse mutations. I also tasted soft, plump mondeuse blanche, the parent of syrah. As we ate Charlemagne's cheese, Grisard pointed down the mountain and shouted, "See that, there? Right there is the only vineyard of mondeuse grise in the world!"

It may have been because it wasn't quite one P.M. and I'd already tasted more than two dozen wines, but looking out over the Combe de Savoie valley, on this unseasonably warm day, a wave of clarity washed over me. I had a vision of a future. Who knows what havoc climate change will wreak on wine grapes. But there's a good chance it will push winemaking farther north and into the mountains, and only native grapes will survive. Perhaps in that cataclysmic future, people will rhapsodize about mondeuse in the same way they now do about pinot noir. If all those future wines taste as good as Grisard's, I'll be more than OK with that. And I'd be deeply grateful for what these wine geeks have saved.

After the tasting with Les Pétavins, we drove north through some of the most famous skiing areas in the world, past Albertville (host of the 1992 Winter Olympics) to the town of Ayse, not far from Mont Blanc and the famed resort Chamonix. In Ayse exist the 54 acres of vineyards where the remaining gringet in the world grows. Half of those acres are worked by Domaine Belluard, one of France's leading biodynamic producers. Belluard is, in fact, an established member of wine's avant-garde, known within the wine-geek bubble, and a darling of the new generation of American sommeliers.

Some believe gringet has existed in the mountainous Savoie since before the Romans arrived. Another legend suggests that, like the debunked story of altesse in Isère and the story of négrette in southwest France, gringet was brought from Cyprus following the Crusades—apparently, French wine drinkers centuries ago loved the romantic idea of their grapes hailing from a mysterious eastern Mediterranean origin. Vouillamoz and his *Wine Grapes* coauthors quote a 19th-century book, *Topographie de tous les vignobles connus* (*Topography of All Known Vineyards*), which made another dubious claim about gringet from Ayse and its effect on drinkers: "This wine has the unique property of not causing inebriation so long as one does not leave the table; but as soon as one takes the fresh air, one loses the use of one's legs, and is forced to sit down."

There had been great anticipation and excitement all day among the Wine Mosaic team about Dominique Belluard's gringet. Etievent, Petit Verdot, and Grisard had talked about it since the morning in Montmélian. But with all the wine we'd been enjoying, we'd fallen a couple of hours behind schedule. The sun was setting and we needed to be in Chamonix soon for a dinner. We arrived at Domaine Belluard amid a lot of buzzing activity. The last of the day's harvest was arriving to be sorted and pressed, while another group seemed to be having a drinking party. Clouds of cigarette smoke hung thick through the tasting area of the winery. Belluard sat at a table scattered with empty bottles along with some friends, a Parisian chef who was apparently of some renown and two Parisian women with whom he was vacationing.

At first, there seemed to be some tension, quite likely because we were so late. Belluard mostly ignored our group. Meanwhile, his Parisian friends criticized a magazine photographer, who was trying to shoot Belluard's portrait, for not shooting in a particular angle of natural light, even though it was almost dark. While we waited, Petit Verdot and I discussed another winemaker's technique of macerating white wines with their grape skins, and Belluard—suddenly engaged and eavesdropping—stood up from the table and shouted at us, slurring his words, "Skin contact! No! No skin contact! No!" Belluard was tall and wiry, with a few days of menacing stubble, and didn't look like the kind of dude I wanted to mess with.

Finally, Belluard poured us a glass of his altesse wine from a magnum bottle. "I think this altesse will be best in another four or five years," he told us. Petit Verdot was in heaven. "I love these wines."

Belluard was now swaying around the winery, popping open bottles—possibly proving that the 19th-century claim by *Topographie de tous les vignobles connus* about losing the use of one's legs after drinking gringet might actually be true. He opened a still wine made from his oldest-vine gringet, which was fleshy, crunchy, spicy, pebbly, and possibly even a little peanutty. The gringet was a strange wine—a bad boy who was also a heartthrob—and Petit Verdot swooned over it like a schoolgirl with a crush: "I feel like there is some sort of unknown stone in this," he said. "This is very special."

Etievent seemed a little embarrassed at how tipsy Belluard seemed, and suggested he take us on a tour of his winery. We followed him into a room filled with a half-dozen concrete eggs. Belluard started to explain that he, like a lot of modern winemakers, preferred these concrete eggs to oak. The concrete softened the wine, but it was neutral and didn't impart oak's influence, which is anathema to younger wine enthusiasts. Belluard claims, as do other like-minded winemakers, that the oval shape of the egg creates a natural vortex, allowing the wine to age on the lees, or dead yeast, without aggressive stirring.

After a few minutes of trying to explain the concrete eggs, Belluard seem to get bored, and he wandered back out to the tasting area. He popped open a bottle of his sparkling gringet, made in the *méthode traditionnelle*, similar to Champagne. "This is better than even the best grand cru Champange!" Bellaurd shouted. While his shouting and demeanor were alarming, it was hard to disagree. This sparkling gringet—smoky, nutty, chalky, razor-sharp—was amazing.

A few days later, after our stop in Switzerland, Etievent, Petit Verdot, and I drove into the French-inflected region of Italy called Vallée d'Aoste, or Valle d'Aosta if you prefer Italian. We traversed the winding road through the Great St. Bernard Pass, 8,000 feet above sea level, one of the most ancient routes over the Alps. It's named for an 11th-century hospice, where the monks famously used enormous Saint Bernard dogs to make Alpine rescues. In popular cartoons and movies, the dogs wear barrels around their necks, filled with brandy, which avalanche victims supposedly drank to stay warm. I vividly remember one Bugs Bunny cartoon I watched as a kid, in which a Saint Bernard pulls Yosemite Sam out of an avalanche, then shakes up a martini, with a twist and an olive, from his barrel. This may go without saying, but the dogs wearing brandy barrels, while a fun story, is complete fiction. Still, we passed a number of cheesy signs at cafés and shops with barrel-wearing dogs.

Compared to sunny Valais, the October weather was cold, gray, and damp on the Italian side of the mountains. Mountainous borders create strange collisions of cultures. In Valle d'Aosta both Italian and French are spoken, along with a local patois. The better-known wineries I visited, such as Grosjean and Didier Gerbelle, and the main appellations, Tourette and Enfer d'Arvier, had French names.

In Villair de Quart, near the town of Aosta, we visited a grower and winemaker with the very Italian name of Giulio

Moriondo who produces under the label ViniRari. We met Moriondo—a tall and slender 57-year-old with lots of curly, graying hair—in his driveway. Waiting outside his garage, where they would be crushed and fermented, sat containers of just-picked cornalin (aka humagne rouge, not be confused with the cornalin du Valais, aka rouge du pays, that I drank in Zürich). Moriondo had taken off work at his day job to meet us: He's a full-time science teacher in the local secondary school. With his hoodie and sneakers, and mellow-but-serious demeanor, he reminded me of my high school track coach.

Just outside the village, surrounded by mountains, we quietly surveyed Moriondo's vineyards, munching random grapes from rows of astonishingly rare varieties with names like petit rouge, vien de nus, fumin, prié, and mayolet. Moriondo said he'd found a new grape this year, a white he called Blanc Common, literally "common white." "It just kept showing up in the vineyard," he said. "No one knows what it tastes like. This will be the first year we make a wine with it."

As we wandered, spitting seeds, Moriondo told us about the various pests he continually has to deal with: badgers, mice, foxes, a Japanese fly that lays its eggs inside the grapes and destroys them. He had set up nets in an attempt to protect the vines, but his body language was that of resignation.

He showed us rows of petit rouge that, until six years ago, had been planted with 80-year-old pinot noir vines. He'd ripped out all the pinot noir. "I don't like the French grapes, the international grapes," he said. "I want to grow only local grapes."

"Come look at this over here," Moriondo said, holding a bunch of grapes that he'd picked from a vine of petit rouge. "This is a new mutation." This bunch of grapes was clearly white. Petit rouge—"little red"—is obviously a red grape. So what was this? A new mutation called . . . Petit Rouge *Blanche*?

Etievent and Petit Verdot became very excited, almost childlike, bouncing up and down. "A new grape variety!" Etievent shouted, with a huge smile. "This is fantastic! You can call it the Giulio variety!" Later, Etievent would tell me, "Grapes are always mutating. We just don't always find them. Or know how to recognize them. But we are always learning."

Over lunch at a local ristorante called Aux Routiers, we tasted Moriondo's stunning wines. Even though he only produces 3,000 bottles a year, more than 20 percent are exported to the United States, where you will find them on the most cutting-edge wine lists. The one that struck me most deeply was a bottle labeled Souches Mères ("Mother Vines") made from vines planted in 1906. Moriondo told us the blend was 60 percent petit rouge, 30 percent vien de nus, 10 percent "cornalin, fumin, primetta . . ." and then a shrug.

I've always maintained that wine is not art. My main reason for feeling this way is that, no matter how great the wine, I've almost never encountered one that conveys complex emotions like fear or loss or grief in the way a great painting or piece of music can. But in Moriondo's Souches Mères, in one of the few times in my life, I felt a profound *sadness* in the wine. I jotted down very few notes that day. Sure, I could tell you that there were autumnal and forest notes and a spicy, smoky character. But there was more to it than that. A few months later, I read wine writer and importer Terry Theise's description of this exact same wine, Souches Mères. Theise wrote: "I sensed the smell of chimney smoke, which alongside the sweet leaf decay comprised a sort of portrait of the eerie yet comforting *tristesse* of this season, the moment before the first snows, the last breath of ruddiness and gold." He added: "Death and beauty often walk arm in arm, after all." While a little more baroque than I might put it, I couldn't agree with Theise more.

Souches Mères also did what all great wines do: it elevated the food, which already was very good, to something memorable. After a week of French, Swiss, and French Swiss food, it was great to be in Italy again, eating thinly sliced lardo, tagliatelle with wild boar, ravioli stuffed with pork and fontina, and a venison stew served with polenta.

The chef, Leo, came out to the table, and we gave him our regards and poured him a glass of wine. Leo told us that his daughter, who'd been serving us, had Moriondo as a teacher in school. "How did it work out?" asked Etievent.

"Ha, nothing! Nothing! She's still here, living with me! I should kill this guy!" He shook his hand in that Italian gesture that means *are you kidding me*? "No, no, I am joking," he said, and we all laughed.

But afterward, it made me feel sort of sad for Moriondo. Even though he is one of greatest rescuers and curators of indigenous varieties in the Alps—a Saint Bernard of lost grapes—he still has to spend every day teaching 15-to-19-year-olds about geology and biology. Maybe if he had just grown and bottled the pinot noir, instead of ripping it out to plant local grapes, he'd have been able to quit his day job by now.

I asked him if he ever brought his students on a field trip into his vineyards. Surely that would be an amazing experience in living biology, right? He shook his head no. He'd never taken his students into the vineyards. "I know it would be great," he said. "But at school, it's always very formal. The focus is always on the classroom and exams."

After lunch, back at the car, Etievent shared with me that Wine Mosaic is worried about Moriondo's vineyards in the long term. Moriondo had no one he was mentoring, and no heir to pass on the knowledge. "He did an amazing job of preserving these varieties," Etievent said. "But he is alone here. The work stops with him."

CHAPTER FIVE

Is Prosecco a Place or a Grape?

WHEN I ARRIVED AT THE ZÜRICH AIRPORT, I found my flight was delayed for several hours. I killed the first half hour at a café over a glass of chasselas. Then I aimlessly browsed the duty-free shops, idly wandering aisles full of overpriced Cognacs, jumbo-size men's fragrances, and the familiar triangular prisms of Toblerone chocolate. It would still be a year before Toblerone controversially changed its chocolate shape, adding more gap space between its classic triangle peaks—a decision that caused Toblerone aficionados to lose their minds, including one Scottish member of parliament who called for government action, suggesting Toblerone was "emblematic of the devastating consequences that Brexit could bring." I guess I had a similarly overheated reaction when I wandered into the duty-free section that specialized in luxury foods such as caviar, salmon filets, foie gras, and expensive Champagnes. Before a display of pricey truffles and truffle oils and truffle-flecked pasta, I began to have an anxiety attack.

Not too long ago, I blew $140 on a half-pound of truffles because of a quixotic desire to re-create a traditional *strangozzi al tartufo*, which I'd memorably eaten on a reporting assignment in Umbria. Like all trips at this stage of my life, I had been in Umbria to taste a relatively obscure wine, a very old grape called sagrantino, named from the Latin for "sacred," that grows on about 900 acres surrounding the hilltop village of Montefalco. Sagrantino di Montefalco may or may not have been described by the first century Roman naturalist Pliny the Elder, who called the grape itriola. Or else it may have been brought to Montefalco by Byzantine monks in the Middle Ages. In any case, by the 1970s, sagrantino was nearly extinct. But—in a story that would become familiar in late 20th-century Italy—a forward-thinking winegrower named Arnaldo Caprai saw sagrantino's potential and saved the local grape. By 1992, Montefalco Sagrantino was promoted to Italy's most prestigious level of appellation.

On my second night in Montefalco at a restaurant near my hotel, I'd eaten a mind-blowing rendition of *strangozzi al tartufo*, paired with the inky, meaty, earthy sagrantino. Sagrantino can be a beast of a wine, possibly the most tannic wine in the world. But when it's good, usually with some aging to mellow it, sagrantino can be as good as Barolo or any other Serious Wine. To this day, I still think about the pairing of Montefalco Sagrantino and that *strangozzi al tartufo*.

Early the next morning, headed to a winery appointment, I told my taxi driver about my dinner. He informed me that he'd likely gathered the very truffle I'd eaten. In fact, he said he was working on no sleep, since he'd been out all night with his brother-in-law and their dogs, combing secret spots in the dark for the precious fungi. This driver, Stefano, disrupted the day's itinerary with a wonderful digression, to a raucous house party of locals, all of whom had brought jugs of their own homemade wine, which we drank out of plastic cups.

When I returned home, I told my buddy Pete about my experience with *strangozzi al tartufo*, and he was intrigued. Pete is a second-generation artisan pasta maker who sells his pasta at Whole Foods, and he hand-cut his version of the "shoelace" *strangozzi* beautifully. "Do you want me to call my taxi-driver friend?" I asked. Pete told me he had a guy who could make the truffles happen.

We faced a dilemma right from the jump. Did we want to buy real Umbrian black truffles at $350 per pound, or could we live with black truffles from Burgundy at $280 per pound? I said: "If my palate is so finely attuned that I can tell the difference between truffles from France and truffles from Umbria, then I've got big problems." We agreed to order the cheaper truffles.

When they arrived, I immediately became nervous that my *strangozzi al tartufo* would never live up to the authentic dish in my memory. The ingredients are simple: olive oil, garlic, anchovies, truffles, salt, pepper, and pasta. Yet just like everything else in Italian cuisine, there are a hundred tiny variations. Would I use anchovy paste or anchovy fillets mashed into a paste with mortar and pestle? Would I mash the garlic with those anchovy fillets or not? What order would I add the ingredients? How would I shave the truffles, and when would I add them? My anxiety quickly grew.

When I prepared the *strangozzi al tartufo* for my family, the first attempt was an instant hit, especially with my sons, Wes and Sander, who scarfed it down with big smiles and approving slurps. I, however, was totally dissatisfied. Yes, the dish tasted great, but something pernicious nagged at me: My *strangozzi al tartufo* didn't taste like what I fondly remembered from Montefalco. Distressed, I quickly prepared the dish a second time, using anchovy paste instead of mashed fillets. I grated another one of my very expensive golf-ball-size truffles, and served it. Still not quite right, I went back to the anchovy fillets, but grated twice as much truffle onto the dish. I was basically keeping my family captive at the table with a truffle grater. By the third plate, Wes and Sander revolted: "We can't eat any more pasta! Please make it stop!"

The dish, in all three variations, had been tasty. Yet I was frustrated and unhappy. What happened? In Umbria, I remembered drinking the dark, tannic sagrantino with the pungent, bold-flavored *strangozzi al tartufo*. My home version, on the other hand, had much subtler truffle flavor. Was I doing something wrong? Maybe, I thought, it was the wine pairing. I was drinking a relatively banal Tuscan sangiovese because I couldn't find a good bottle of Montefalco Sagrantino. Montefalco Sagrantino is the sort of wine that always finds itself sort of in-between: At around $35 or more per bottle, it's too expensive for everyday drinking, yet it doesn't have the special-occasion cachet of the Serious Wines. And sadly, it's not the type of wine that gets love and support from many of the too-cool young sommeliers, who believe it's too big and oaky.

Then I had another, more sinking feeling: Maybe I could taste that we'd purchased the cheaper truffles. A metaphysical chill ran down my spine. Could I discern a Burgundian truffle from an Umbrian truffle? Was it bad enough that I could definitely tell the difference between a Tuesday-night bottle of bland Tuscan red and a fine Montefalco Sagrantino? To what end was all this tasting? Was all of this very helpful to my life? I am a perpetually one-step-ahead-of-broke writer who should be much more frugal with my money than I usually am. No, at that moment, at my sink scraping off the remains of $140 worth of truffles, it seemed that this knowledge would actually make my life less enjoyable, and certainly less sustainable.

I'd been experiencing the same nagging, sickly feeling lately about my journey into wine. Every so often, there's some kind of populist study showing that most "normal" people (read: not super-fussy wine snobs) cannot really tell the difference between, say, $9.99 bottles and $100 bottles of wine. Sure, I'd done a great deal of self-education and put in lots of trial-and-error, and I fancied myself one of those people who could *usually* discern between a cheap, mass-produced wine and one from a quality producer. But could I really, all the time, with 100 percent accuracy? When I'd read the research behind those populist "gotcha" blind tastings, I guessed probably not. But beyond the imposter syndrome lay something deeper, even more disquieting. Taste takes a commitment, and practice, just like yoga or golf or Tantric sex. For years I've been on some kind of elusive quest of taste. But has this quest into pleasure led toward some enlightenment or happiness, or has it simply succeeded in making me a miserable person?

I occasionally worry about these sorts of things. I am well aware how ridiculous or pathetic that may sound, the ultimate First World Problem. Taped above my desk is a yellowing *New Yorker* cartoon from several years ago: A chic-looking man and woman sit at a table and gaze at one another over glasses of wine; the woman, her hand clutching at her bosom, says to the man, "Do wine writers suffer and all that?"

At the Zürich airport, I still had a few more hours before my flight to Italy, and so I decided to feel fancy and cashed in some miles to relax in the airline lounge. I'm not normally a lounge kind of guy, and I felt a little self-conscious in my gray hoodie and sneakers, among the slim businessmen, in their sharp blue suits and pointy shoes. On the television, Donald Trump was shouting at a campaign rally, and I felt like the people in the lounge were staring at me, the slovenly American, through their cool, Euro-designed eyewear. I filled up a bowl with gummy bears and then a wine glass with some cheap prosecco from a magnum bottle sitting in an ice bucket. The prosecco was a little lukewarm and a little sweet, but it paired well enough with the gummy bears and a shrug. As Eric Asimov, the *New York Times*'s wine critic, once wrote: "You cannot brood over a prosecco."

Of course, prosecco has become so ubiquitous—the fuel of *Real Housewives* drama, book clubs, and girls-night-out pregaming, as well as the bubbly ingredient in trendy aperitif cocktails like the Aperol Spritz and Bellini—that I sometimes almost forget that it's a wine at all. Because it's inexpensive (usually $12–15) and enjoyed in a casual, by-the-glass atmosphere, prosecco is almost always referred to by serious wine critics as "relaxed" or "breezy" or "perfect for summer drinking." Translation: Serious Wine people do not take this wine too seriously.

It's perhaps now hard to believe, but just a decade or so ago, prosecco was still mostly overlooked, while other sparkling wines took the spotlight. It wasn't until the late 2000s that it truly started blowing up. Between 2007 and 2009, the amount of prosecco consumed by Americans grew more than 65 percent. This was during the worldwide financial crisis, and one can surmise that people were looking for a cheaper, easier alternative to expensive, serious Champagne. The amount of the grape planted in Italy rose almost 60 percent from 2000 to 2010, to more than 48,000 acres. Prosecco was so much in demand that the grape started to be grown in places other than the Veneto, throughout the rest of Italy and even other countries. You could find prosecco from California and Australia and Brazil.

During this time, Paris Hilton appeared on *Late Show with David Letterman* to promote the launch of her new sparkling wine, called Rich Prosecco, which was sold in cans, with an expiration date, for around $3. "And prosecco?" Letterman asked Paris. "What does that mean?"

"It's Italian," she said. "It's like Italian champagne."

"Italian champagne. In a can? Champagne in a can?" asked Dave.

"It's sexy," Paris said. "It looks great when you're holding it."

"Sexy?" Dave chortled. He then proceeded to shake the can of prosecco to see if it would explode. "Oh, you're right. That is sexy."

Thus, a vast portion of the American public was introduced to the charms of prosecco. But Paris Hilton's canned prosecco didn't sit well back home in prosecco's historic homeland, the region of the Veneto flanked by the towns of Valdobbiadene and Conegliano. In fact, Paris Hilton's canned prosecco drove them into a rage—most especially the fact that the grapes hadn't even been grown in Italy, but in *Austria* for God's sake!

In March of 2009, the Italian agricultural minister Luca Zaia called the rise of foreign prosecco "agropiracy" that threatened "the future of Prosecco" and "hurts the perception of the Made in Italy brand." It became a major political issue for Zaia (a member of the right-wing, nationalist, anti-immigrant *Lega Nord* party), who eventually was elected president of the Veneto.

The Italians' solution for protecting prosecco? Create a new, expanded *Denominazione di Origine Controllata* (or Controlled Denomination of Origin, abbreviated as DOC), enforced by the power of the European Union. This response was nothing new. In the recent past, Italians have very publicly tried to police everything from fake prosciutto di Parma to inauthentic Italian restaurants abroad to the proper way to make Neapolitan pizza. And it's not just Italy. The French likely also cringed when Paris and Dave talked about prosecco as "Italian champagne." When it comes to sparkling wine, the government-funded Champagne Bureau spends a lot of money and effort lobbying to make damn sure we all know that "Champagne" only refers to sparkling wine produced in the Champagne region of northern France. In fact, nearly every wine region has protected and controlled denominations: In France it's *Appellation d'Origine Contrôlée* (or AOC); in Spain, it's *Denominación de Origen* (DO); and in Portugal, it's *Denominação de Origem Controlada* (DOC). At this point, Europe has hundreds of protected designations of origin, with France and Italy having more than 300 each.

There was an obstacle, however. Prosecco was a grape variety, and not a geographical term. The European Union only protects products originating from specific geographic regions. Legally, the prosecco grape could be grown anywhere, in the same way that you can grow chardonnay or pinot noir in Burgundy, South Africa, or New Jersey. This is where the Italians got intensely creative. They located a village in Friuli, near Trieste, named Prosecco. And then they simply renamed the grape. They called it glera. No matter that the Friulian village of Prosecco wasn't really known for its sparkling wines. No matter that no one in the Veneto ever really called the grape glera. The authorities carefully drew DOC boundaries so that they included the village of Prosecco, and then the consortium of prosecco makers told everyone that the "ancient"—"Roman" perhaps?—name of the grape was *glera*. *Voilà*! As of the 2009 vintage, Prosecco was now a place and not a grape. The sparkling wine made in this new geographic construct, Prosecco, would now be protected by law, and no one outside of northeastern Italy could legally call their sparkling wines "prosecco."

Ampelographers and grape experts like José Vouillamoz were not so enamored by this decision. In *Wine Grapes* (where the grape is still listed as "PROSECCO") the authors call the name change "confusing and misleading" and point out that glera was actually a slang name for a less common clonal variation called prosecco lungo, which has oval rather than round grapes. This wine-geek opposition fell on deaf ears.

Further, an even stricter denomination was created for the collection of towns stretching from Conegliano, a half hour north of Venice, toward Valdobbiadene, through the Venetian hills an hour to the west. The wine from this area is now known by the tongue-twister name Conegliano Valdobbiadene DOCG Prosecco Superiore—DOCG meaning a Controlled and *Guaranteed* Denomination of Origin, Italy's most prestigious appellation. In other words, with just a few jujitsu moves of government, Prosecco protected its commercial interests and positioned itself as a Serious Wine. Not everyone thought this was a good thing. As food writer Alan Richman wrote, in his *GQ* column, in 2011: "The Italians, because they're Italian, have made Prosecco more complicated than it needs to be." The formerly humble, inexpensive, overlooked sparkling wine was now a diva.

With 377 known grape varieties grown throughout the Boot, it is an understatement to say that wine in Italy is complicated. Even its most famous wines can be famously maddening. Nebbiolo, the grape of Barolo and Barbaresco, is called picoutener in the alpine Valle d'Aosta; spanna in Novara, in Piedmont not too far from Milan; and chiavennasca in Valtellina, in northern Lombardia, near the Swiss border. Most wine drinkers know that the important grape of Tuscany is sangiovese, from which Chianti is made. But sangiovese has nearly 50 synonyms. In the town of Montepulciano, for instance, sangiovese is called prugnolo gentile. (For maxiumum confusion, there is also a grape variety called montepulciano that is not related to sangiovese.) Meanwhile, the most well-known sangiovese synonym is brunello, from which the elegant and expressive Brunello di Montalcino, one of the most prized wines in the world, is produced around the village of Montalcino. But drive just over the mountains toward the coast, and into the wild, southernmost region of Tuscany called Maremma, and sangiovese here is called morellino. Why is it called morellino? Some think it comes from the word *morello* (meaning "brown") which supposedly refers to the color of the native horses in Maremma. Why would anyone name a wine grape like this? Who knows? Perhaps as we proceed, we can all agree that delving into Italian wine grape nomenclature is to descend into madness.

In 2009, I was in the Veneto doing research for my first book, for a chapter on Italian aperitif cocktails—such as the

Aperol Spritz and the Negroni Sbagliato, both of which call for prosecco as an ingredient. By happenstance, I was invited to a lunch at the castle of Borgoluce winery, to celebrate the new DOCG designation. Borgoluce, which produces prosecco as well as cabernet sauvignon, merlot, and pinot grigio, is a huge winery, with over 3,000 acres of vineyards. Before lunch, the Borgoluce public relations person whispered in my ear, "The countess would like to sit next to you." Well, let it be said that I am not one to disappoint a countess. The countess, Maria Trinidad di Collalto was, of course, an extremely charming woman of an indeterminate age. She seemed slightly older than me, and yet looked a decade younger. We were joined at the table by a couple of junior writers from an important American wine magazine, and some of the countess's friends from Bologna. The table next to us was full of journalists from Korea, the next one with journalists from Eastern Europe, and someone identified as "the most famous sommelier in Italy" and host of a popular TV show on wine.

"This is a very exciting time for prosecco, a revolution, a big adventure," said the countess.

One of the young American wine journalists pressed the countess. "Don't you think we have to reeducate the public?" he said.

"But people love prosecco," the countess said. "I love prosecco."

"But many people think that prosecco is just prosecco," the journalist said.

The countess smiled and leaned in close in an endearing, conspiratorial way. "We take ourselves too seriously sometimes," she said.

Later, we were invited to a dinner at a Veneto restaurant called Da Lino. The journalists from the American wine magazine were there. We were also joined by several prosecco producers, the director of the Conegliano Valdobbiadene consortium, and their PR person, named Silvia. There were journalists from Estonia and Poland, and one who wrote for an influential Austrian wine magazine called *Falstaff*, and who looked like Paul Shaffer, David Letterman's old musical sidekick. Dinner began awkwardly when the Polish journalist loudly complained that the PR people had taken him out for sushi the night before. "Sushi! In Italy!"

Paris Hilton and her canned prosecco came up before the first course. The Austrian journalist who looked like Paul Shaffer insisted that Paris Hilton was good for prosecco. "Look at how many stories were written about this controversy," he said. "Look at how many more people know the word prosecco now."

But the Prosecco people were worried. The sparkling wine market is tricky in the United States and that is clearly where they saw the future growth of exports. Champagne was already a household word, and cava, the inexpensive sparkling wine from Catalonia, such as Freixenet was saturating the market. "Do Americans even know that if they buy cava, that it comes from Spain?" asked Borgoluce's winemaker.

"Yes and no," said one of the journalists from the American wine magazine. "By now most wine buyers vaguely know that champagne comes from France, cava from Spain, prosecco from Italy." But he added: "Most don't get more specific than that."

"Would Americans be drawn to a bottle with the words Conegliano Valdobbiadene?" asked Silvia. "Do you think this would be something, would mean quality, to the American consumer?"

"I don't think so," I said.

"No," said the American wine journalist.

"What if we called it Prosecco *Superiore*?" asked Silvia. "Superiore, like they do in Chianti and other areas?"

"Possibly," I said.

The Austrian journalist who looked like Paul Shaffer didn't like the name Prosecco Superiore at all. "Nonsense!"

"Look," said the young American wine journalist. "You have to understand the psychology of a certain type of American wine consumer. He or she is terrified. Terrified that he or she is going to bring the 'wrong' bottle to a party or whatever. On the shelf there are all these foreign names that are impossible to pronounce. If you have two bottles of prosecco, and one says 'Prosecco' and the other says 'Prosecco Superiore' and that bottle costs a few dollars more, that person will buy the Superiore. It's driven by a sense of fear."

I nodded my head in agreement at the time. But now I think he was a little harsh. I mean, you have to give us Americans a little credit. We do at least know that sparkling wine should always come in a bottle.

There was still at least an hour before my flight would board, and my prosecco and gummy bears were gone, so I went back up the buffet and bar area of the lounge. The food spread was the same as you find in anonymous lounges and hotel breakfast rooms all over the world: vegetable soup, lunch meats, sliced bread to be heated in an industrial toaster, goopy pasta salad, wilted greens, green tomato slices, croissants, potato chips, M&Ms, trail mix. I scouted around for wines made from the interesting Swiss grape varieties, but there were only the usual suspects: pinot noir, tempranillo, malbec, chardonnay. I poured another glass of prosecco; it was a little cooler, but still a tad sweet. I looked at the label and saw that the wine classified as *Extra Dry* (in English). Just like naming of grapes in Italy, labeling wine sweetness levels is also convoluted. Italian sparkling wines labeled *Dry* (again, in English) actually have the highest sugar content, with about 20 to 26 grams per liter. Extra Dry usually clocks in at about 14 to 20 grams of sugar per liter. *Brut* is actually the driest, with anywhere from six to 12 grams of sugar per liter. Extra Dry has long been the most popular for Italy and its sweet tooth, and is generally the kind you'll be poured during *aperitivo* hour. Nearly every good producer I've met, however, expressed a preference for brut.

I was happy enough with the Extra Dry, which I paired with some trail mix and potato chips. I was grateful to be able to afford this brief interlude of luxury before returning to Economy class. More than that, I was grateful that I was able to pursue this crazy passion at all. Over prosecco and trail mix and chips, I meditated on where my obsession with obscure wine grapes and this journey had begun. I could clearly pinpoint my awakening.

It was spring of 2010, and my first book, on spirits and cocktails, was about to be published. I was having a pretty good run as the *Washington Post*'s drinks columnist. I had loyal readers who, in sincere emails and social media posts, regularly asked for my advice, I was invited to speak at events and conferences, and every bartender I met seemed to have a special cocktail just for me. For once in my life, people looked to me like some sort of expert on some topic, not just a wayfaring generalist, as I'd always been. The problem was that once I finished my book, I was bored with spirits and cocktails. I was totally ambivalent about my so-called expertise, and of being seen as an expert. I felt similarly to cultural critic Geoff Dyer, who in his essay "My Life as a Gatecrasher," wrote: "If I'd known what I needed to know before writing the book, I would have had no interest in doing so."

And so, as I restlessly awaited my book's publication in the spring of 2010, I relentlessly lobbied and pestered my editor to let me write about a new topic: wine. In my naïve hubris,

I assumed that wine was just another drink, just like whiskey or brandy or Chartreuse or a pre-Prohibition cocktail, and since I was already being told I was an expert on those drinks, I assumed my supposed expertise would magically leap to this new beverage topic: wine. As I mentioned earlier, I had no idea how wrong I was. Further complicating my plan was that the newspaper already had a regular wine columnist. So this presented a significant obstacle for me, a fraught issue for my editor, and a tense situation for our writer-editor relationship. But I was insufferable, and so I ended up taking a wine assignment with another editor at another newspaper, with the misguided idea of forcing my editor's hand.

In any case, this is how I ended up in northeastern Italy again, to report on some crazy developments that were happening with prosecco. In retrospect, I realize that the main reason I was given this assignment was because I'd already written about prosecco as an ingredient in cocktails. Also, because no proper wine writer still took prosecco seriously.

My trip was supposed to be a four-day jaunt. The plan: jet into Venice; visit a dozen top prosecco producers; jet out; return home; write article. I focused on the Serious Prosecco, in the Conegliano Valdobbiadene DOCG bull's-eye. Even the growers there were still getting used to the new nomenclature. "My grandfather never called it glera," said Franco Adami, then the president of the Conegliano Valdobbiadene consortium, with a chuckle. "But you have people all over the world who want to plant quote-unquote *prosecco* so we have to protect it."

I found noteworthy in my travels within the DOCG that prosecco production is still a relatively young phenomenon. Even until the mid–20th century, most prosecco (or so-called glera) was produced as a still wine, with the sparkling wines only gaining popularity in the 1960s and 1970s. "I really thank Paris Hilton," said Matteo Bisol, third generation of the Bisol prosecco family. "That was when everyone here realized we had a problem. Without Paris Hilton, we wouldn't have a DOCG."

I was amazed at how many tiny vineyards were planted on incredibly steep slopes in the foothills of the Alps. "Valdobbiadene is not an area, like Tuscany, with large vineyards," said Antonio Bisol, Matteo's grandfather. "Here, these two rows may be owned by one grower, and then these two rows over here are someone else's." Nowhere is this truer than in the "grand cru" of Conegliano Valdobbiadene, the Cartizze, a steep hill with only 106 hectares. At two million euros an acre, Cartizze rivals Montalcino, where Brunello is grown, as some of the most expensive vineyard land in all of Italy.

"When I speak with winemakers in other parts of the world, they say, 'Ah it's not a complex wine. It's a simple wine.' But this is not true at all," said Stefano Gava, the young winemaker at Villa Sandi. "This is not a forgiving grape like chardonnay. Making glera, making sparkling wine, is difficult. This is not a full-bodied grape. It's so delicate. If you make one mistake, you have a big problem."

Most astonishing to me was just how legitimately, consistently better the DOCG proseccos were when compared to most non-DOCG we normally find in the United States. You'll end up spending about $5 to $7 more for a bottle from Conegliano Valdobbiadene, but if you do you'll usually notice an immediate difference—less cloying, crisper flavors, and more elegant aromas than what's generally poured by the glass. I also realized, of course, how difficult it would be to tell this story to an audience that's grown accustomed to paying $11.99 a bottle for something they consider a fun wine.

Yet I tasted some significantly aged prosecco, 1990 and 2000 vintages, when I visited Primo Franco, winemaker at Nino Franco winery. As fine prosecco ages a decade or two, it takes on an attractive amber color and develops a honeyed nose with a beautiful, long finish—yet there's still a little hint of effervescence left. The 1990 Nino Franco, in particular, was remarkable. On that beautiful afternoon, we also sipped his Nino Franco 2008 Grave di Stecca bottling (which retails in the United States for over $40). Despite these serious bottlings, Franco expressly wanted to make one thing clear: "Always remember, prosecco is good when it is easy to drink."

After the deep dive into prosecco, I was ready fly home. But I soon learned I wasn't going home. On the last day of my trip, an Icelandic volcano with the unpronounceable name of Eyjafjallajokull erupted, spewing tons of ash and causing havoc for air travel. Many, at the time, called Eyjafjallajökull the worst disruption in the history of transportation. Like millions of others during that shutdown of European airspace, I hadn't factored a volcano into my plans. The airline canceled my Sunday morning flight home from Venice, with no possibility of returning before at least Thursday. I'd be stranded in Italy for several days.

Poor Jason! Stuck for four extra days in Italy! As you can imagine, there was very little sympathy forthcoming from family, friends, and coworkers when I texted them the news. "Awwww. It must be SUCH a struggle to be stranded in that boutique hotel featured in *Architectural Digest*!"

"You can always get a boat home," texted my friend Pete, the pasta maker. "That's how my family got over to the States."

When I let one friend know of my predicament, she simply texted: "You suck."

Indeed, life can be full of struggles. But perhaps this was not one of them. On Sunday, the first day of my exile, I accompanied Matteo Bisol to lunch at the restaurant his family had just opened on Mazzorbo, one of Venice's outer lagoon islands. It was a warm, sunny day, and we ate outside across from a small vineyard, dating to the 14th century, where the Bisols were attempting to rescue an ancient Venetian grape called dorona. We ate these delicious soft-shell crabs you can only find in Venice. And also consumed lots of prosecco.

"Everything OK?" my mother texted.

"Yes," I wrote, "All is fine, I'm just boarding the vaporetto back from lunch, and Matteo is going to give me a tour of Venice's wine bars."

No further reply or concern from Mom.

I'd love to tell you of a single hardship. That I paid some insane thousand-euro fare for a taxi to take me to an open airport, where I had to sleep on a cot. That my boss was really upset with me. That my children forgot who I was. But no. I basically just spent several more days drinking wine and eating in Italy.

At a certain point, I felt like the overprivileged son of a deposed dictator, one who lives in the lap of luxury, and yet will never go back to his homeland. The night before my flight was canceled, I'd had dinner in the beautiful hill town of Asolo (where they also have their own DOCG, Asolani Prosecco Superiore). A famous exile, Caterina Cornaro, the queen of Cyprus from 1474 to 1489, was sent to Asolo after she may or may not have poisoned her husband. She lost Cyprus and was named Lady of tiny, fairy-tale Asolo to "rule" as consolation. During Cornaro's exile, the Italian verb *asolare*—meaning to pass time in a delightful but meaningless way—came into usage. Perhaps that's how I can sum up my brief stranding in Italy. I visited some more wineries. Made some more friends. *Sono asolato*.

I often fear that this sense of asolare permeates my life. Chuck Klosterman, in his book *Sex, Drugs, and Cocoa Puffs*, delineated "no-nonsense guys" and "all-nonsense guys," identifying with the latter. After years on the wine-and-spirits beat, it's not always easy to make the case that I'm not an "all-nonsense guy." But I know I'm not the only one who believes that wine and food can offer profound experiences. Like Caterina Cornaro holding court in her make-believe domain, perhaps we're kidding ourselves that Umbrian truffles make a better pasta dish than Burgundian ones, or that one should waste money on truffles at all. Maybe it's delusional to see politics in the shape of a Toblerone chocolate. Perhaps it's an indulgence of the mind to think that DOCG prosecco is much better than non-DOCG prosecco. But these are weak moments of doubt. As old Kant told us: *Happiness is not an ideal of reason, but of imagination.* As old Nietzsche told us: *All of life is a dispute over taste and tasting.*

During my exile, I used my time to visit wine producers outside of the prosecco region. On the last day, I paid a visit to Cesari, one of the best-known producers of Valpolicella wines. "You can't drive Nature," Deborah Cesari told me. "Nature drives you." She was talking about making Amarone della Valpolicella, the full-bodied, high-alcohol, brooding Serious Wine made from grapes dried for months on straw mats to concentrate their juice. Cesari explained how her family is only able to release vintages from certain vineyards in certain years, and how the grapes determine this, as well as if the wine will be aged for months or years, and whether that aging will happen in large barrels or small. "We never know what Nature gives us." Of course I hear some version of this trope every time I ever visit a winery, no matter where it is. (*Great wine is made in the vineyard, not the cellar.*) But during those days I was stuck in Italy because of that Icelandic volcano, the idea that Nature drives the world took on special meaning.

After the Amarone, I spent another sunny lunchtime *asolare* at a restaurant facing Lago di Garda. With my poached pike, I drank a bright, mineral-driven white wine from Lugana, a tiny region along the lake, made from a grape called turbiana. I found this simple bottle of turbiana even more interesting than the huge, muscular Amarone I'd tasted earlier, which would have overpowered the delicate lake fish for lunch. But the idea of liking the turbiana more than Amarone made me uncomfortable—I knew enough about wine to think that I *should* like the big, brooding Serious Wine better than the local Lugana.

A little while later, after I began seeking out Lugana wines at home, I would learn that the people around Lago di Garda once believed that turbiana was a local variation of trebbiano, one of the most widely planted grapes in the world. But DNA testing in 2008 confirmed that it was actually verdicchio. At the time, I hadn't quite gone down the rabbit hole of esoteric and rare grapes, and so turbiana (or verdicchio) from Lugana still seemed fresh and unknown to me, and the idea that ampelographers were still uncovering new knowledge about it was incredibly exciting. Not too much later, I saw the news that the region was under a severe threat. Due to the proposed construction of a high-speed bullet train, Lugana would lose 750 acres of vineyards, more than a quarter of its vines. #SaveLugana soon became a popular social media hashtag among wine people. My seeking out and drinking Lugana began to feel even more important.

"Yet what should become of the world?" D. H. Lawrence writes in his classic travel book, *Twilight in Italy*, a series of sketches about his stay in Lago di Garda from fall of 1912 to the spring of 1913. "[T]he industrial countries spreading like a blackness over all the world, horrible, in the end destructive. And the Garda was so lovely under the sky of sunshine, it was intolerable." Lawrence uses the slow peasant existence of Lago di Garda as a metaphor for all that is good and pure in the world, setting it against what he calls the "purpose stinking in it all, the mechanising, the perfect mechanising of human life."

I was actually reading *Twilight in Italy* during my stranding in Italy. Lawrence's book is strange, as is all great travel writing. But *Twilight in Italy* is also something else that all great travel writing seems to be: prescient. On the surface, Lawrence's account is about old women spinning wool or lemon gardens or ancient churches. But at the same time, it's undeniably about war, the collapse of the old order, the coming fascism, "mechanising of human life." Lawrence left Lago di Garda only slightly more than a year before Austrian Archduke Franz Ferdinand was assassinated in Sarajevo, leading swiftly to the outbreak of World War I. The book's publication only predates Mussolini's rise to power by six years.

That night I went to my dinner at Dolada, a Michelin-starred restaurant in foothills of the Alps, overlooking the serene Lago di Santa Croce in the village of Pieve D'Alpago. I went to Dolada with a husband and wife who happened to be rival prosecco producers (and two of the finest), Cinzia Canzian of Le Vigne di Alice and Umberto Cosmo of Bellenda. If anyone in the Veneto is pushing the boundaries of what prosecco can be, it is Canzian and Cosmo. These two are not making prosecco to be used in an Aperol Spritz.

A major reason that the wine world looks down upon prosecco is that, unlike sparkling wines like Champagne, prosecco uses the Charmat method, in which the secondary fermentation takes place in a stainless steel autoclave rather than in the bottle. Both Bellenda and Le Vigne di Alice have been experimenting with bottle fermentation, or *metodo classico* prosecco. Canzian has a number of bottlings with zero dosage, or no sugar at all. Cosmo has a bottling he calls *metodo rurale*, with little intervention and no sulfites. "Nobody has tried to push the limit of what glera can be," Cosmo said. "We are not even close to the limit."

Before dinner at Dolada was served, Cinzia and Umberto debated what was more "serious," red wine or white wine. "All of the wines I would call 'unforgettable' are white wines," said Canzian. "Certainly I drink a lot of wonderful red wines. But for me, white wines are much more often surprising and memorable."

Cosmo protested, but Canzian's argument was tough to refute since the white wine we were drinking at that moment, the one Umberto had ordered, was totally unforgettable: Vigneti Massa Timorasso Derthona. This was a strange, complex, full-bodied white, one that felt like everything, all at once, honey, ripe fruit, freshness, minerality, a unique nuttiness, and more than a little funk. After phylloxera, timorasso had dwindled to less than 20 acres around the town of Tortona, in the Piemontese province of Alessandria. But in the early 1990s, a winemaker in Tortona named Walter Massa nurtured the grape back from the brink of extinction.

To pair with the timorasso, Cosmo urged me to order Dolada's *nuovi* spaghetti alla carbonara, chef Riccardo De Prá's special take on the humble bachelor's dish of pasta, eggs, and bacon that De Prá serves "deconstructed." De Prá told us, "It's not about a recipe. It's about a concept." The carbonara came served with a soft-cooked egg hidden under a coating of ground pepper, atop a nest of beautiful golden pasta and crispy guanciale. In order to reconstruct what the chef had deconstructed, we broke the yolks and tossed everything together before tucking in. The richness of the pasta, and the mingling sensations of crunchy, peppery, and creamy was unbelievable. This was the greatest rendition of carbonara I'd ever eaten. Yet the experience was doubly amplified by the timorasso in the glass.

Cosmo mused sentimentally on how he used to make carbonara as a young man at university, and something about the coziness of this memory, paired with the foreignness of the wine and my odd week of exile, made me miss home badly. But I can pinpoint that day as my awakening. First with my lunchtime discovery of Lugana, followed by dinner with avant-garde winemakers who were pushing prosecco, or glera, to its limits, and finally the weird, profound, ancient timorasso.

This was the moment I began my quest to discover rare grapes, my journey into obscurity, and my attempts to bring others along with me. I felt like the thirsty voice in Emily Dickinson's poem:

I bring an unaccustomed wine
To lips long parching
Next to mine,
And summon them to drink;

Sitting in the airport lounge in Zürich, years later, emptying the glass of middling prosecco, I yearned for my next unaccustomed wine.

CHAPTER SIX

When Wine Talk Gets Weird

WE LIVE IN A GILDED, rococo age of lifestyle advice. Wine—much like home decor, cooking, fashion, or grooming—is one of the mysterious spheres of cultural knowledge that service journalists are forever trying to "demystify" or "simplify" or "hack." There's no stopping the torrent of wine advice that spews forth: from columns, blogs, Twitter feeds, YouTube videos, morning news shows, apps, and books; from so-called wine educators promising to make wine "more accessible"; from sommeliers who utilize "education" to sell more bottles. "Knowing a little something" about wine occupies a more prominent mindspace to a certain aspirational segment of upper-middle-class Americans than does knowing a little something about contemporary art or foreign languages or how their local municipal government works.

Yet when I think about us all drowning in a sea of wine tips, I think back to a shark attack that happened more than a decade ago at the Jersey Shore. A teenage surfer had his foot bitten by what, experts believed, was a baby great white shark. Luckily, even though he received 60 stitches, the boy survived and made a full recovery. Now, shark attacks in New Jersey are rare. Exceedingly rare. This was actually the first attack in decades. Our local newspaper ran a front-page article about the incident. The headline: "For Some, Ocean Loses Appeal After Shark Attack." The subhead: "But others not as concerned and take the plunge into Atlantic."

Alongside that article ran a box of helpful tips under the title "WHAT TO DO: If you come near a shark." Tip #1: "Don't try to touch it." Tip #2: "Get out of the water as quickly as possible." Tip #5: "If a shark attacks you, the general rule is do whatever it takes to get away." Now, there's some service journalism for you! But is it any less helpful than so much of the wine advice we get? Not that all of this advice is bad, though much of it is self-evident or common sense or unnecessary. It's helpful in the same way that, when you lose your keys, it's helpful for someone to ask, "Now, where did you last leave them?"

"Wine is the only thing that makes us happy as adults for no reason," said Saul Steinberg, as quoted by Adam Gopnik in his book *The Table Comes First*. Yet, as Gopnik points out, we find so many ways to make ourselves unhappy over wine. To which I would add: stressed or confused or frustrated. This wine anxiety has spawned a cottage industry of wine education and paraphernalia.

When I began to learn about wine in earnest, for instance, I purchased a kit called Le Nez Du Vin that professed to teach me how to identify various aromas in a glass of wine. The kit, which was imported from France, came in a dictionary-size case covered in red fabric so that it resembles an old book. Inside are a dozen tiny glass vials, each of which is redolent of a specific, essential red-wine scent when uncapped. These vials are cosseted in crushed velvet (or likely velour). It was purchased at Williams-Sonoma. It cost $130.

Le Nez Du Vin contains two slim manuals, both written by Jean Lenoir, a French wine critic who almost 30 years ago developed this method of wine education by way of aromas. In the first book, Lenoir lays out his methodology, explaining the primary, secondary, and tertiary aromas in wine. He talks about fruity notes like black currant and cherry, floral notes like rose and violet, vegetal notes like green pepper and truffle, roasted notes like smoke and dark chocolate, and animal notes like leather and musk. He explains how the sense of smell works and how it relates to the "art" of wine tasting. He also explains, with a diagram, that I should not ever drink the liquid in the tiny vials (as well as keep them away from my skin and eyes) and that I should do my wine-aroma education in a quiet room, "free of extraneous odor such as tobacco or perfume."

In Lenoir's second manual, he explains the aromas in each of the 12 vials—strawberry, raspberry, black currant, cherry, violet, green pepper, truffle, licorice, vanilla, black pepper, and "smoked"—in detail. (These 12 are particular to red wine—there is also a white-wine kit, presumably for another $130.) Lenoir lays out how some selection or combination of these 12 primary aromas can be found in the main grape varietals and wine-making regions. Sangiovese will have aromas of strawberry, raspberry, licorice, and smoke. Pinot noir will have cherry and violet, black currant, and licorice. Bordeaux and Burgundy will exhibit "smoked" notes. Lenoir suggests that if I practice sniffing the 12 vials in his kit hard enough, I can learn to identify these 12 aromas by memory. Thus, I'll be able to identify just about any wine by smell. "Concentrate on your olfactory perception," Lenoir writes. "You detect an aroma? Can you name it? . . . [L]et the thoughts and images come flooding into your mind, even some emotionally charged memories tied to certain moments in your personal history."

I have to admit that the first few times I used it, I was skeptical: *Look at me, the fool who flushed $130 down the toilet!* But now I'm pretty happy I acquired Le Nez Du Vin. "We are taught to read and write and count, why not to smell?" Lenoir asks in his manual. Yes, he's just sold me a $130 kit about sniffing wine, but Lenoir insists he is driving at something deeper, that aromas can "open the door to your own private scent memories." A whiff is enough to send you right back to childhood. "There you are once again," he writes, "standing in that wheat field just after the harvest; or you might be breathing in the baking smells of your grandmother's kitchen and there she is in her apron, smiling down at you. It brings it all back."

We all, of course, know this to be true. Smell brings back the most elusive and yet most visceral memories. The emotional memories we have almost no language for. And it's rarely a smell as simple and straightforward as an apple pie at grandma's. For me, the smell of Aqua Net can stop me cold, not only taking me back 20 years ago to specific big-hair high school crushes that went unrequited, but also dredging up the indescribable feelings of wanting to run far, far away from my southern New Jersey hometown. A whiff of a certain patchouli oil reminds me of a beautiful, hippie, dippy college girlfriend in Vermont who broke up with me because I made fun of her Saturday-night drum circle at the Environmental House, but it also conjures up an odd sense of regret that my adult life has little space in it for patchouli or hippie-dippiness. Sniffing a handful of potatoes snaps me all the way back to bittersweet childhood memories, of walking into the cool refrigerated warehouses during hot summers at a packinghouse once owned by my father. The emotions of that particular scent are so tightly wound, they may never be unraveled.

OK, I realize this is the sort of touchy-feely stuff that makes people roll their eyes over the whole wine thing. What does a $130 wine-sniffing kit have anything to do with all this, right? Well, on the most basic level, you can't learn how to taste until you learn how to smell. Jancis Robinson, one of the world's foremost wine critics, writes: "Until studying wine in my 20s, I had not been taught to use my sense of smell, so I am well aware of how underused it can be." Robinson once described in the *Financial Times* a "terrifying" period when she'd lost her sense of smell due to a strange flu. "I was all for telling the world my career was over," she wrote.

Further, you can't learn how to smell unless you're willing to spend time sniffing, followed by free association. It's one thing to sniff and identify aromas of "cherry" or "vanilla" or "licorice" or whatever. But so what? There are not literally cherries or vanilla or licorice in the wine. Our mind is simply trying to process the sensory experience of those smells or tastes, in the same way we do when we hear the sounds of music. There's another step after actually smelling: our mind free associating to create some kind of meaning. As the philosopher John Dilworth suggests, in an essay in the anthology *Wine & Philosophy: A Symposium on Thinking and Drinking*, "wine is only the raw material for a series of highly personal improvisational experiences."

Le Nez Du Vin may be helpful in training the nose and the mind. But for most people, the step of free association is a difficult one. As Dilworth suggests: "Most people are too inhibited to think of themselves as being capable of engaging in any artistic-like activity, let alone of a kind that requires them to freely and creatively extemporize a personal performance or interpretation of something." Thank God, then, that wine offers a quick solution to lower those inhibitions: alcohol. "The alcoholic content of the wine provides a kind of permission, or entry ticket, into a parallel world in which—in the terminology of Immanuel Kant—*a free play of the imagination* can take place," writes Dilworth. This is why, when I teach a wine class, the room is silent when we taste the first wine, but by the time I pour the fifth, everyone is shouting their deepest thoughts on what they're tasting.

OK, so when I start quoting philosophers, I realize that I'm probably going overboard and into that sketchy, semi-ridiculous space that makes normal people think wine geeks are full of shit. So let me stick to the important question about Le Nez Du Vin: Does the damn kit work or not? At the risk of sounding inscrutable: I think this is too personal a question to answer. I've done numerous experiments with friends and students who are not professionals, and few have been able to guess all 12 aromas in the kit. People identify strawberry as "floral," truffle as "mushroom," blackberry as "grassy." No American I know has ever guessed "black currant" correctly, since it's not a common taste we grow up with, unlike in Britain. I find that a lot of confusion comes from people trying to match up the kit's smells with fashionable wine-tasting terms that are often thrown around by critics and sommeliers. It's hard work to honestly put language to smell, especially when our heads are spinning with so many buzzwords.

When I first acquired Le Nez Du Vin, I had two subjects who didn't have any wine-tasting language floating around in their heads. Two people who were completely, blissfully ignorant of wine knowledge. These would be my sons, Sander and Wes, who were only six and four years old at the time. Though it still would be many years before they would have a sip of wine, I'd decided to see if they could guess the smells in the kit's vials.

First up was Sander. I opened up vial #30, Green Pepper, and waved it under his nose. "It smells like salad," he said. Salad, incidentally, was not a good smell to Sander. OK, I thought, we're sort of on the right track.

I opened vial #12, Strawberry, and let him sniff. "This one smells like fruit snack," he said.

With a sinking feeling, I opened vial #13, Raspberry. "This one smells like strawberry fruit snack," he said. OK, so Sander was not on his way to sommelier school just yet.

Next up was his little brother, Wes. I opened what I thought was a difficult one, #54, "Smoked"—which smells like bacon or other smoked meat. Wes sniffed and said, "This smells like salami."

I was impressed. Did we have a savant on our hands? Well, I thought—patting myself on the back—tasting does run in the family. Maybe the boy's a genius? Maybe he could become the Doogie Howser of the wine world? When could we sign him up for the Master of Wine program?

I opened vial #30, Green Pepper, the same one Sander had sniffed. I put it under Wes's nose.

"Here, Wes," I said, beaming, "tell me what you think this is."

"This smells like Jell-O," he said.

Sigh. Oh well.

The next time I used Le Nez Du Vin was a number of years later, when I was asked to design a wine course at the university where I was working, to be offered in the International Area Studies department, called The Geography of Wine. By that time, my spirits book had been published, I was now regularly publishing articles on wine, and I spent numerous evenings hosting wine tastings or casual classes. But I'd never been faced with teaching a full ten-week academic term on wine. The people I usually taught about wine were also older professionals. I'd never had the responsibility to shape the wine knowledge of a class full of 21-year-olds. I spent much of the spring and summer of that year pondering how I would do such a thing. What the hell would I teach these kids about wines?

I knew I wasn't the only one searching for how to connect young people to wine. Earlier that winter, I'd received a press release and some samples for a "new, edgy" brand of wines "targeted at millennial consumers" called TXT Cellars. The line included the following four bottles:

- OMG!!! Chardonnay
- LOL!!! Riesling
- LMAO!!! Pinot Grigio
- WTF!!! Pinot Noir

I am not making this up. The marketing material talked about how these wines were "unpretentious" and "easy to relate to" and the tasting notes avoided "wine geek talk" because "We don't like wine geeks." According to the press release: "It is a wine that isn't meant to swirl around in your glass and sniff but to enjoy with friends and 'LOL.'" Ahem.

TXT Cellars certainly wasn't the only wine company targeting young drinkers. Cutesy labels like Cupcake Vineyards or Middle Sister Wines or Be Wine (featuring Flirty Pink Moscato and Radiant Riesling) or the multicolored store displays of the HobNob line or the girly bottles made no secret as to their target demographic.

But it isn't just cutesy brands. Among serious wine professionals, I hear a ton of chatter about marketing wine to millennials. Many believe this enormous generation of people in their 20s and 30s will be the saviors of the wine industry. The market research seems to back it up. Almost concurrently with the launch of TXT Cellars, the Wine Market Council released a survey showing that millennials most closely mirror what the Council terms "high-end wine buyers" (meaning people of all ages who buy over-$20 bottles at least once a month). Like high-end wine buyers, they are more likely to consult wine reviews, more likely to visit wine bars than either Gen Xers or Baby Boomers, consume more wine per occasion, and of course spend more time on social media.

The most encouraging sign, from the Wine Market Council study, may be that millennials are more likely to try wines from grapes they've never heard of before. Many people who sell wines from lesser-known grapes seem to have pinned their hopes on millennial wine consumers and their eclectic tastes in grapes.

I spoke with a cool millennial branding consultant in Los Angeles named Leah Hennessy, who had blogged about the Wine Market Council's survey. I called her one Saturday afternoon, and asked her to tell me what the kids were drinking. Hennessy's interpretation was straightforward and succinct: "Millennials are drinking more wine and better wine at a young age than any other generation has." This is all very exciting for people in the industry who hope that millennials will lead an adventurous new age of wine consumption, caring about both quality and value, championing rare and obscure grapes like blaufränkisch and savagnin and mondeuse and chasselas and gringet instead of old standbys like chardonnay and pinot grigio and cabernet sauvignon, and challenging, once and for all, the stuffy wine Establishment.

Still, when I hear this high-minded talk of millennial wine drinkers, I often think about the university students I used teach in creative writing courses. Lots of my students had studied abroad and were way more educated about, with more fully formed opinions on, what they eat and drink than those in my generation ever were at the same age. However, another segment of my millennial students also told me all about something called Slap the Bag. This is a drinking game in which the bag of wine is removed from, say, a box of Franzia. The bag is held high, and everyone slaps the bag while someone chugs from the spout—the harder the slap, the bigger the chug. Now, I make no judgments. But my point here is that the true wine-drinking nature of a generation as vast as the millennials likely falls somewhere between the educated, adventurous, high-end wine buyer . . . and those who play Slap the Bag.

"Each generation has a different approach to wine," Hennessy told me. "A lot of millennials drink wine as a social accessory. It says something about us. There's a mystique, it's mature. You're not drinking beer, which already says something about you."

One important finding that the Wine Market Council's survey verified is that wine labels matter, a lot. Millennials were twice as likely to cite the importance of a "fun and contemporary wine label" than Baby Boomers. "If your label is too traditional, it's off-putting," Hennessy said.

Hennessy spoke of the importance of "gateway wines," or wines that hook young people and cause them to seek out more. "The whole point of a gateway wine is to feel empowered to try something else. Is it interesting in some way? Is the label interesting? Is there a story?" she said.

While I was working on the syllabus for The Geography of Wine, I assembled a half-dozen millennials for a small blind tasting, with the bottles covered in paper bags. I included some of the cutesy brands such as Cupcake and HobNob, as well as a selection of what I might consider "gateway wines" from Europe—mostly $8 and under reds from Spain and Portugal.

When the paper bags were removed, it was clear just how absolutely important the labels were. Innovative graphics and bottle designs were given high marks, such as the slender blue bottle and sans serif font of Relax Riesling ("A perfect party wine"). Another favorite was Grooner!—a cheap grüner veltliner labeled with its phonetic spelling and exclamation point, colorful illustrations, and copy that read "Perfect for Parties . . . Great with Food . . . Picnics too!" The millennials were even more impressed with the modern labels for the reds from Spain and Portugal I had assembled. They preferred the bright orange labels of Borsao garnacha or the spare, black Monte Velho from Herdade do Esporão in Portugal's Alentejo—neither of which overtly screamed "millennial!!!" "This is what I want a wine bottle to look like," said 22-year-old Alex, a business student. "It's aspirational. It feels mature."

In the tasting, I also included the TXT Cellars wines like LMAO!!! Pinot Grigio and OMG!!! Chardonnay. Among the group, these came in for the harshest criticism. "Ohhh noooo, I hate this so much," said Kinsey, a 21-year-old who worked at an ad agency. "I'm embarrassed that this is what they think people my age want."

"This conveys like partying or drunkeness to me," Alex said. "I drink wine for the same reason a 50-year-old woman does. If I'm going to get drunk at a party, I'll just drink vodka." I didn't have the heart to tell him that partying and drunkeness are why plenty of 50-year-old women drink wine.

Right around the time I was struggling over the syllabus for The Geography of Wine, and becoming a little bit frustrated, I found myself standing out in front of a liquor store in downtown Philadelphia, at noon on a Wednesday. Please don't misunderstand: I wasn't sucking down a bottle wrapped in a paper bag. I was sober and still gainfully employed. Actually, I was standing there as part of my job as a wine journalist. A half-dozen others stood with me, and they too seemed gainfully employed, or at least employable. "I'm supposed to be in a meeting right now," said a guy with white hair and trim mustache wearing a pastel golf shirt.

Believe it or not, our little group was patiently waiting to be taught how to navigate the aisles of Pennsylvania Fine Wine & Good Spirits store. We all held little laminated maps of Spain that highlighted its major wine regions, and we waited, maps in hand, to be led on what our instructor, Michael McCaulley, called a Wine Safari.

Safari, of course, conjures up images of machetes and pith helmets and gun bearers and dangerous wild animals. It may seem a bit extreme to equate wine with big game, but perhaps it makes sense. So many people I talk to seem to experience wine shopping with the kind of panicked confusion of someone lost in the Serengeti. They speak as if the chance of hunting down an interesting, good-value bottle is as likely as bagging a rare, albino elephant (or something like that, just to torture the safari metaphor as far as it will go).

McCaulley is the managing partner of Tria, a group of excellent wine bars in Philadelphia, the kind of places where drinkers might first taste off-the-beaten-path grapes like a zippy gros manseng from southwest France, a rosé zweigelt from Burgenland, a funky mencía from Ribeira Sacra, or a cool teroldego from Trentino. He also directs the Tria Fermentation School, which teaches a full calendar of wine classes. In other words, he is a locus of wine education in a city like Philadelphia.

The lunchtime Wine Safaris were meant to be half-hour primers and the advice was meant to be shorthand, or a refresher for people who have already taken classes at Tria. McCaulley, for instance, would toss out some advice like, "Look for the three Vs: vermentino, verdicchio, and vernaccia. If you find a grape from Italy that begins with the letter *V*, you'll get a good white."

On that day, McCaulley led us through the Spain and Portugal shelves. "When we talk about Spanish whites, we often talk about albariño," he began, grabbing a bottle of white wine from the section clearly labeled SPAIN. "Albariño is from Galicia in the northwest, or 'Green Spain.' It's from an area called Rías Baixas, which has the highest per capita consumption of seafood in Europe. So obviously people there drink these wines with seafood." There were lots of head nods from the group, though no one except me was taking notes.

Over the next 25 minutes, he led us, whirlwind-style, from Spanish whites like verdejo and godello, to reds like tempranillo from Rioja and Ribera del Duero (labeled "Denominación de Origen, or D.O." on our tiny maps) and grenache (we were told it's called "garnacha" in Spanish), and then on over to the Portuguese shelf where he talked about popular, under-$10 vinho verde ("fresh, and tends to have a little spritz") and under-$20 Portuguese "country reds" from Dão and the

Alentejo. He told us that in Portugal "tempranillo is called 'tinta roriz.'" Then he showed us sparkling cava from Catalonia ("which is not champagne") and then moved to crisp, dry fino and manzanilla sherries that we can pick up for $15.99. "Sherry right now is going through a renaissance," he said. Without a pause, McCaulley then jumped over to bottles of Blandy's five-year-old and ten-year-old Madeira. "Madeira is a Portuguese island off the coast of Africa," he says. "My favorite wine in the world is Madeira."

With that, the Safari ended. Perhaps everyone's brain was spinning as much as mine, because there were no questions from anyone in our group. The guy in the golf shirt returned to his meeting. Two people left their little maps behind in the store. Alone, I puttered around by the Spanish shelves for a bit, pondering the Wine Safari. McCaulley had laid a lot of information on all of us, and I wondered what parts of it people would recall later. Mostly, though, I marveled at the fact that this group of aspirational, professional adults felt enough anxiety to think they needed someone to coach them on how to buy a bottle of wine—but also enough curiosity to seek out the advice. It seems impossible that these people hadn't previously read or heard some kind of wine advice in some form.

For instance, if you click on the "Wine Pairing" category on Amazon, there are no less than 414 books available, all purporting to solve the ultimate First World Problem of what to drink with what you eat. Yet how much of this hand-wringing is useful or relevant once we learn of a recent industry study, which found more than 60 percent of wine consumed by "high-frequency wine drinkers" is consumed without a meal? Understand that we're talking about real wine drinkers, too, not just the person who pops open a liter of Barefoot or Skinny Girl or Yellow Tail once a year at a holiday party. "High-frequency wine drinkers"—according to Wine Opinions, the market research firm that conducted the study—refers to the 29 million wine drinkers who consume the beverage at least several times per week. These drinkers drive more than 80 percent of the wine market, and almost all of the wine is over $15 per bottle. If these kinds of people are ignoring wine pairing advice, then what's the point?

I knew I didn't want to teach wine by demystifying or simplifying or providing life hacks. Wine is not a simple subject, and I can totally understand the desire for shortcuts. But I'm often approached by people who casually tell me they want to learn a little something about wine. I'll tentatively ask them if they are sure, and they'll say yes. Then, inevitably, the moment I start trying to explain a basic point about a grape, or where those grapes come from, or how the wine is aged, or whatever, they just tune out. To me, that's like saying you want to understand baseball, but you refuse to learn what the shortstop does or how a curveball is different than a fastball. With wine, effort will always be rewarded. I fervently believe that wine can be experienced on a deep cultural level. Perhaps not exactly like art or music or literature, but close. Just like any other cultural activity known to humankind—bridge, bird-watching, bowling, BDSM—learning more about the topic opens up a whole new world of experience.

At the same time, I also acknowledge this basic truth: You don't have to know anything about wine to drink it and enjoy it. All you really have to know is how to operate a corkscrew, and even then you could just stick with a screw cap, or boxed wine, or just push the cork in. Further, knowing something about wine doesn't make you more "sophisticated" or "cultured" or "classy." It doesn't raise your IQ. It doesn't increase your penis size. I always chuckle when I remember a 2007 study that claimed 22 percent of men in the UK embellished their wine knowledge to impress dates.

After the Wine Safari that day in downtown Philadelphia, I hung around the store and chatted with one of the sales clerks, a tall, skinny guy with a beard and glasses named Max Gottesfeld, to whom McCaulley introduced me as follows: "Max is a crazy man who likes wacky wine."

Since Prohibition ended in 1933, the state has controlled all wine and liquor sales in Pennsylvania. Max is one of the state's official "retail wine specialists," a position requiring advanced training that was created by the Pennsylvania Liquor Control Board for employees who sell premium wines. Oddly enough, that position was only created in 2012—eight decades after Prohibition ended. Why did it take so long to commit to employees that have actual wine knowledge? Well, back in 1933, Governor Gifford Pinchot claimed the purpose of the Liquor Control Board was to "discourage the purchase of alcoholic beverages by making it as inconvenient and expensive as possible."

While things have changed (slightly) since then, sometimes it seems our wine education and knowledge is still just as conflicted as it was coming out of Prohibition. In the classic post-Prohibition bartending guide *The Official Mixer's Manual*, written in 1934 by mixologist Patrick Gavin Duffy, there is a chapter called "A Guide to Wine Drinking," by Murdock Pemberton (an early writer at the *New Yorker*). In it, Pemberton wrote:

> *Another factor, which delayed the spreading of the wine habit was the aura that sprang up around wine, confusing honest folk and misleading the timid who might otherwise have become wine drinkers. Too many of us were frightened away by the needless and careless advice that wine drinking required special equipment, that wine drinkers must belong to a certain cult with a difficult ritual . . .*
>
> *Then, to offset this high-toned approach, many dealers began a backfire of educational advertising. In this they often went so*

> *far in the other direction that their claims were ridiculous and certainly as harmful to a spread of wine drinking as the cultured talk of old experts.*

Pemberton was writing at a time when wine ran a distant third to cocktails and beer, and the number of wines imported into the United States was much smaller. His list accompanying the chapter included Bordeaux, Burgundy, Champagne, German riesling from Rheingau and Mosel, and sherry, port, Tokaji, and Madeira. No Italian wines. No Spanish wines. No Swiss wines. No wines from South America or South Africa or Australia. Not even a mention of California or other American wines except a passing mention that the quality was nowhere near that of France. Fast forward to the 21st century, and there's never been a moment in history when you could drink more wines from more diverse places. Yet so much of traditional wine education still focuses on Bordeaux, Burgundy, and Champagne. Meanwhile, other formerly Serious Wines like port or sherry or Madeira or German riesling are sadly misunderstood. Perhaps this is why we need something like a safari guide in the wine shop, someone to help us venture beyond the overpriced and boring, beyond the usual suspects. We need a new map. Or perhaps a sherpa, perhaps in the form of a skinny, bespectacled wine clerk named Max, "a crazy man who likes wacky wine."

"There's been more of an investment in esoteric wines here," Max told me, as he grabbed bottles of furmint from Hungary, treixadura from Galicia, monica from Sardinia. A twinkle came into his eyes as he enthused about wines from far corners of the world or from obscure grapes that most consumers aren't familiar with: xarel-lo from Catalonia, mourèvdre from Provence, ribolla gialla from Friuli. To be sure, Max was totally geeking out. But Max wasn't being obscure for the sake of obscurity. What he was most excited to point out was how inexpensive so many of these wines were. "If you go slightly off the beaten path," he said, "that is where you find good value."

I realized I wanted to serve as a similar sort of sherpa. Learning about wine is not something that happens in a night, in a weekend, or in a ten-week academic term. You can simply never learn everything there is to learn about wine. Wine is as big as the world itself. How could I teach a bunch of kids about the world? All we could maybe do during a single term was to begin the journey.

"That's kinda gross," said one of my students, a young woman who'd come to class on Halloween dressed as a bottle of malbec ("vintage 1990").

I'd just rubbed my thumb across the chalkboard and licked it, in a vain attempt to explain what I meant by the "chalky" finish of the Sancerre we had just tasted—this in comparison to the grapefruit-and-cat-pee New Zealand sauvignon blanc we were also tasting.

"Cat pee? Ew," said Malbec's roommate, who was dressed as a bottle of cabernet sauvignon, and who'd previously told me that she'd bought Cupcake Vineyards' Red Velvet wine because she was "excited to know if it would actually taste like a red velvet cupcake."

As we tasted our next two wines, someone stopped me when I'd inadvertently used the au courant wine-geek term "minerality" to describe the Chablis we tasted in comparison to the Napa Valley chardonnay.

"Wait," said the wrestler who sat in the back row. "Minerality?"

"Think of the sensation of licking stones."

He looked at me blankly. "Do you go around licking stones?"

"Kinda gross," said one of the communications majors.

"No, no," I pleaded. "Picture licking pretty, smooth stones. Not dirty ones."

I further compounded my floundering explanation by uttering what must be the worst word in the wine lexicon: *mouthfeel*. Let me be clear: You simply cannot say mouthfeel in a college classroom and expect 20 students—and their professor—to keep a straight face. I knew this because the week before I'd made a similar mistake, referring to a wine's "creamy mouthfeel."

Later, after we'd finally gotten to the reds, I opened a big, monstrous Gigondas with a bit of that telltale Rhône aroma. I poured it and waited. There were lots of scrunched-up noses. "Ugh, funky," said the wrestler. "What is that smell?"

I hesitated . . . and then asked, "Does anyone get a hint of what wine people call ummmm . . . a barnyard, or farmyard, aroma?"

"Barnyard! You mean like cow manure?"

"Well, maybe farmyard is a nicer way to put it?" I said.

"Ew," said the girl dressed like cabernet sauvignon who'd hoped a wine tasted like red velvet cupcakes, with her arms crossed. "That is totally gross."

Please don't misunderstand. It wasn't the wines that my students found gross. My students liked the wines, loved some of them, even the Gigondas with the farmyard aroma. No, it was the descriptions, the standard wine-world terms, that turned them off.

Overall, it was an amazing, inspiring experience to teach the university course, like having a front-row seat to observe how young people experience wine for the first time. It was amazing to watch their knowledge grow exponentially week by week. But if there was one stumbling block, it's when we left the comforting aromas and flavors of fruits and flowers and herbs and entered into more challenging tasting territory: Minerality.

Chalk. Tar. Tobacco. Animal. Farmyard. Petrol. "Why would we want to drink a wine that tastes like these things?" my students wanted to know.

It's a reasonable and valid question. Look, I tell them, if you're happy and content with fruity, pleasurable red wines redolent of berries and cherries and plums or zippy, easy-to-drink whites with tangy citrus and orchards full of apples and pears . . . well, then that's what you should drink without feeling any need to move beyond that. Wine should be, foremost, about pleasure—and pleasure is personal. There's a reason that romantic comedies with happy endings, sunny, catchy top-40 music, mac 'n' cheese, flavored vodka, and wearing Ugg boots with pajama pants remain popular.

But if we think more deeply about pleasure, we realize it isn't always so straightforward or even comfortable. After all, why do so many of us love sad poems, disturbing horror films, or intense psychological dramas? Why am I capable of loving dark albums like Bruce Springsteen's *Nebraska* or the Smiths' *Meat Is Murder* or Elliott Smith's *From a Basement on the Hill*—while at the same time I can enjoy summer pop hits like Luis Fonsi's "Despacito," Daft Punk's "Get Lucky," or Drake's "One Dance."

With the arts, we inherently understand that without the darker, more confounding elements, there can be no light. Wine is no different. Just as in novels or films or musical compositions, the more complex and ambitious the wine, the more unique and potentially discomforting aromas, textures, and flavors we'll find.

One crucial step for the novice wine drinker is to move beyond just fruit and to embrace the concept of minerality. What is minerality? For me it takes many shapes and forms. Is it like wet stones? Chalk? Slate? Flint? Talcum? Shells? Like water from a well or a cistern? If I'm being really expressive, it reminds me of a certain bracing drink of cold, fresh water that I once took from a rocky pond in an Icelandic glacial meadow, where I was led in the midnight sun by a beautiful Icelandic woman. Um . . . yeah . . . see how hard it is to pinpoint this thing called minerality?

So much ink has been spilled on the subject of minerality, some of the best discussions launched by fine colleagues such as Jordan Mackay ("the unknowable . . . is the most alluring taste of all"), Alder Yarrow ("Pundits argue the meaning of minerality, but what escapes definition may still be tasted."), and Steve Heimoff ("I can't define 'minerality,' but I know it when I feel it."). Clark Smith, in an article for *Wines & Vines*, writes that "minerality is not an aroma, nor is it a flavor by mouth" but rather "an energetic buzz in the wine's finish, almost like an electrical current running through the throat."

Beyond minerality, other hard-to-define elements happen because of aging. Wine is, after all, a living thing, and as the years pass, magic happens inside the bottle as fruity and floral qualities transform into something more savory. That's why we find a well-aged riesling that smells like gasoline, a decade-old Brunello di Montalcino with aromas of old leather, a cellared Barolo wafting hints of tar and asphalt. But it's not always aging. I've tasted young Bordeaux Superieur and Spanish monastrell with notes of sweet chewing tobacco, and a young, earthy Madiran with a funky whiff of the cow pasture.

"Yeah, yeah, OK," I can still hear my students say. "But *why* would I want to drink wines like these?"

It's the kind of controversial question debated by wine professionals. For years, it was understood that classic vintages of some of the world's finest wines, say Château de Beaucastel from Châteauneuf-du-Pape or Penfolds Grange from Australia, would emit a touch of a manure-like farmyard (or barnyard) aroma in the nose. The reason for this is well known: That barnyard smell comes from the brettanomyces yeast—known as "brett" in wine circles. It's the same wild yeast used to make many of the sour beers that have become so popular. While it used to be tolerated much more in Old World wines, influential critics and judges have begun to view brett as a serious flaw in wine.

A similar sort of reconsideration is happening with the "petrol" aromas found in riesling. Now, first of all, the fact that we fussily borrow the British term *petrol*, rather than using the American *gasoline*, shows how uneasy we feel about this aroma and how to describe it. But a good riesling, with *at least a few years of aging*, has wonderful petroleum aromas that range from those little rubber Superballs that I used to play with as a child to the smell of the pumps at a small country gas station on a crisp spring afternoon—and this quality is sought-after by riesling fans.

Since a new generation of sommeliers has been on an evangelical mission to turn us all on to the pleasures of riesling, the enjoyment of petrol notes is becoming more mainstream. But since it's still off-putting to novice drinkers, many winemakers have moved to downplay the word *petrol*. You rarely see it on labels, for instance. The German Wine Institute has omitted mention of petrol in its German-language version of the official Wine Aroma Wheel. Some winemakers, in some cases, have even declared petrol a defect. Famed winemakers like Olivier Humbrecht of Zind-Humbrecht in Alsace have declared within the past couple of years that young rieslings should never smell of petrol. For these winemakers, petrol doesn't begin to show itself until about five years in the bottle.

For me, with all of the wilder, funkier flavors in wine, it's always a question of balance and context. Take the farmyard or barnyard quality. Like many people into craft beer as well as wine, I love Belgian saisons and lambics—which have a "bretty" character—and so a little brett doesn't immediately turn me off. But there is a line, however moveable, that can be crossed.

During the term, a friend who is a prominent wine blogger

visited my wine class to talk about Australian wines. The topic of minerality and petrol came up as we tasted Clare Valley riesling. After class, we took our guest speaker to Tria, McCaulley's wine bar, to sample its diverse list. One of my students, Marco—who eventually became a wine writer—decided to step outside his comfort zone and order a glass of Cantina Sociale Cooperativa Copertino Riserva 2004, made from the negroamaro grape in Puglia. I ordered one too.

When the glasses arrived, I took the first sniff and said, "Animalistic." Marco took a whiff and said, "Wow. Oh. My. God. This is like Sex Panther," referring to the pungent cologne from the Will Ferrell comedy *Anchorman: The Legend of Ron Burgundy*.

He started passing it around to the others. The wrestler leaned in and took a whiff, and said, "It smells like a barn."

"It smells like a horse's butt," said Malbec and Cabernet Sauvignon.

Finally, our wine critic guest took a sip, and told me, with a wink, that he'd consider sending it back—he being of the camp that viewed brett as a flaw. Our server walked by right then, and politely disagreed. "I really like this one," she said. "I love a funky wine like this." For me? Well, at first, I thought it crossed that delicate line from pleasure to bordering on too pungent. But once I let it sit for a moment, and the server brought our stinky cheese and a plate full of cured ham . . . well, we were back once again in the world of pleasure.

Soon enough, I ordered a second glass.

In history as in human life, regret does not bring back a lost moment and a thousand years will not recover something lost in a single hour.

— STEFAN ZWEIG

II.

TRAVELS IN THE LOST EMPIRE OF WINE

CHAPTER SEVEN

Wines with Umlauts

IF THIS WERE AN ALTERNATIVE HISTORY, in which I imagined that the Austro-Hungarian Empire had not collapsed in 1918, then the wines of Austria would not now be considered obscure. In fact, the umlauted wines of the Habsburg monarchy would be rivals to Bordeaux and Burgundy, counterbalancing France's influence on the wine world. Perhaps if the Empire had survived the slaughter of World War I, blaufränkisch, grüner veltliner, and zweigelt might be the noble grapes grown and sought-after in places like Napa Valley, Oregon, and Chile. But this is not an alternative history. The Empire did fall, two World Wars did happen, and Austria's amazing wealth of indigenous wine grapes remains little known and underappreciated.

By the mid-to-late 20th century, Austrian viticulture was in a desultory place. With its borders redrawn, several once-prized wine regions now existed in new countries, some behind the Iron Curtain. The wines continuing to be produced in Austria were largely ignored. Sure, there were still drinkable white wines enjoyed fresh and young at local wine taverns. And the country's dessert wines still had a small following abroad. In fact, the Austrian wine ladder was classified by sweetness level, with the sweetest wines fetching highest prices.

Then, in 1985, disaster struck. In a ham-handed, criminal attempt to increase sweetness levels, a group of unscrupulous wine merchants were caught adulterating their wines with a chemical, diethylene glycol, used in manufacturing antifreeze. The poisoned-wine scandal, reported worldwide (and a few years later spoofed in *The Simpsons*'s first season), led to millions of gallons of Austrian wines being pulled off shelves in most countries, including the United States, which temporarily halted imports. Whatever tiny market Austrian wines had at that point was destroyed. Besides antifreeze-laced wine, 1985 was also the year that alleged Nazi war criminal Kurt Waldheim successfully ran for president, so it was a particularly bad year for Austrian public relations.

With its wine industry in shambles, the Austrian government stepped in and passed some of the strictest wine laws and quality-control procedures in the world. While keeping true to Old World ideals and indigenous grapes, younger winemakers embraced certain New World technologies. They were, for instance, among the first Europeans to embrace screw caps over cork. Sweet wines (which had sometimes been adulterated) were shunned in favor of dry wines. A new emphasis was placed on farming practices, with Austria eventually having the highest number of organic producers per capita in Europe. In short, Austria's entire wine scene hit a big reset button, with younger producers at the forefront. Fred Loimer, a top winemaker in Austria's Kamptal region, told me: "I was a young man in 1987, and I started making wine in the dark days. But I am of the lucky generation. The generation before us made a huge mistake and so they didn't get to have a say in what came next." During the 1990s, Austria transformed itself into one of the most dynamic wine countries in the world. The first sign of this was the rise of its native, tough-to-pronounce grüner veltliner, which by the turn of the 21st century was in demand among sommeliers and in-the-know wine drinkers.

But there are many more wonderfully enigmatic grapes to be discovered in Austria. So I arrived to look into what happens when geopolitics remakes the wine map.

Lake Neusiedl, in Austria's Burgenland region, is a bizarre body of water, ringed by a wide belt of tall reeds and often engulfed by fog. Splashing across the border into Hungary, it is a so-called steppe lake, warm and brackish and gray, with no outflow and a single, tiny tributary that replenishes its evaporation at a trickle. At times, the lake becomes so shallow that you can wade across it. There have even been times in history, such as in the 1860s, when the water has dried up completely.

This region once belonged to the Kingdom of Hungary, and the land around the lake was controlled by the noble Esterházy family, Hungarians who for centuries were loyal to Austria's Habsburg emperors. The Esterházy princes cared very much about arts and culture. Famed composer Joseph Haydn lived and worked at Schloss Esterházy for decades. The Esterházys also cared a lot about wine, planting vineyards of varieties like blaufränkisch, neuburger, furmint, and grüner veltliner near the banks of Lake Neusiedl. It's been said that Haydn often took part of his salary in the form of wine.

But that was all long ago. In 1921, after the fall of the Austro-Hungarian Empire and the Habsburg monarchy, Burgenland became part of Austria. Twenty-five years later, after the Second World War, the nearby Hungarian border was sealed off behind the Iron Curtain. If the Esterházy castle were not still standing in Burgenland, the whole notion of princes dabbling in world-famous wines and music would seem as distant and preposterous as a fairy tale.

I visited Lake Neusiedl at Illmitz, a small town at the southeast shore, where I wandered the vineyards along the lake as a brilliant autumn sun began to set over the span of reeds. A dreamy, eerie silence, broken only by the sounds of a few birds, enveloped the shores.

I was in Illmitz to meet Christian Tschida, an avant-garde, 39-year-old winemaker whose bottles are in demand at some of the world's most cutting-edge restaurants. I'd first tasted Tschida's wines in Copenhagen at several of that city's wildly inventive New Nordic restaurants, including Noma, which at the time was regularly considered "the best restaurant in the world." Tschida was sought after by the younger generation of sommeliers who wanted so-called natural wines—unfiltered, minimal technology, no sulfites added. One of his wines was named Laissez-Faire. The labels themselves were also provocative, some adorned with artist Mel Ramos's cheeky images of voluptuous, nude women. Tschida's home, which is also his winery, sits about 100 meters from the Lake Neusiedl.

Tschida answered his door with disheveled hair, wearing an old Pink Floyd T-shirt. "You've found me on a busy day," he said. "I've been up for 24 hours in the cellar." When I set up my appointment the day before, he'd told me that his success had made him wary of foreign visitors: "We're getting a lot of people dropping by, but the wrong kind of people. They come to the winery because they read about me, but then they say, 'This wine is unfiltered. I don't like it.'"

Before we tasted, Tschida made pour-over coffee and offered me some. "This is the perfect coffee to pair with the wines," he said. I looked at him dubiously. Was he putting me on? For many serious wine professionals, coffee before a tasting would be unthinkable. Wine critics and sommeliers maintain strict rituals and quasi-sterile environments: no brushing teeth before tasting, no perfumes or cologne, no smoking, and definitely no coffee. For whatever it says about me, I accepted a cup of coffee.

Tschida then opened his refrigerator and poured the first wine, a red called Kapitel I. "This is zweigelt and cabernet franc. It's been open for two weeks." I took a gulp of coffee and looked at him, again, to see if he was putting me on. "What do you think of it?" he asked.

I took the glass and swirled it around, took some into my mouth, swished it around, and swallowed. "Um, well," I said. "The aromas and flavors are still popping, but I think it's lost a little body."

"Maybe," Tschida said. "Yes, maybe. But I don't really care about body." He dumped what was in our glass, smiled, then opened a new bottle and poured again. Thus commenced one of the stranger tastings I've been part of. Here's how Tschida, in a translated prose poem on his website, describes Kapitel I, his blend of zweigelt and cabernet franc:

> *"The Dark Side of the Moon." Cool dark aromas around the wine. A wine like an exemption: Rugged means lead to the goal. Always fighting against an excess, an excess of banality. The sweetness curbed with all available means, in favor of unruly aromas. Delicate bitterness of wild arugula, piquant acid, which is painstakingly worked out from the hot, Pannonian plains. Old vines, rooted in the search for food, deep in gravel, under the humus. How else can this unfathomable depth grow?*

I don't know that I can improve upon—or completely make sense of—that description. I can say that Kapitel I, at 12.5 percent alcohol by volume, is light for a red wine, compared to, say, a 14.5 percent Tuscan sangiovese or a 15 percent garnacha from Spain or a 15.5 percent zinfandel from California. This low alcohol is a hallmark of new-generation winemakers. Tschida despises high alcohol. "Around 13 percent you can taste the alcohol. Above that, the alcohol takes over."

Tschida works mostly with local varieties. "Austria is the Garden of Eden," he said. "You have so many different grapes.

You can do whatever you feel like." As I tasted Tschida's strange wines, made with grüner veltiner and blaufränkisch and sämling 88, I wondered how the average drinker back home would react to these labels with their tricky Germanic pronunciations. I recalled an unfortunate truth once put forth by *New York Times* critic Eric Asimov:

> *But nothing scares wine consumers so much as an umlaut. This diacritical mark, a mere two dots, is simply intended to communicate a particular pronunciation of the vowel it tops. Instead, it often signals to consumers that they must flee in terror to the nearest bottle of pinot grigio.*

Regardless, I was particularly taken by an eccentric wine called Himmel auf Erden ("Heaven on Earth"), like a fresh pear cider, garnished with a meyer lemon and peach blossoms, sipped from rustic stone earthenware. Himmel auf Erden is an odd blend of pinot blanc and sämling 88 (translated as "seedling 88"), the latter also known as scheurebe. Scheurebe, which has been having a moment among the wine cognoscenti at home, was created in 1916 by crossing riesling with a wild vine. A prominent importer of Austrian and German wines, Terry Theise describes scheurebe as "like riesling just after it read the *Kama Sutra*." (Again, yes, another risqué simile.)

Anyway, because Himmel auf Erden is so peculiar, Tschida said the Austrian wine authorities forced him to bottle it as *landwein*, or generic-area table wine, and not a *qualitatswein*, or quality wine from a specific region. "It's not typical for our region, and so it's been rejected as a *qualitatswein* three times." One of the reasons is because scheurebe is not an official grape. As a *landwein*, he's not even permitted to put the country's red-and-white flag on the bottle top, standard with most exported

Austrian wine. Yet Himmel auf Erden still commands high prices from certain top restaurants.

Tschida studied fine arts at school, but took over the family vineyards instead of pursuing the life of an artist. As a young man, he made and drank very traditional wines. "I was spending all my money to taste the best wines of Bordeaux, of Burgundy. I was always looking for the perfect wine. I was looking for the Holy Grail," he said. "I did not find it."

In 2004, Tschida decided to go in an entirely new direction. He sold off all his barriques, the smaller barrels that can convey prominent oaky flavors to wines, a style that gained wide consumer popularity in the 1990s and early 2000s. By 2005, he went completely organic in his vineyards and returned to the old practice of foot-stomping his harvest instead of using a pressing machine. In 2006, he began letting his white wines macerate with their grape skins during fermentation (which winemakers generally only do with red wine, in order to give them color). Rather than red or rosé, the contact between the white wine and its grape skins creates so-called "orange" wines that have become all the rage among younger sommeliers. "I love skin contact because you get these amazing aromas. It's like traveling to the center of the earth," he said. Tschida works with wild, native yeasts for fermentation rather than industrial starter yeasts. And he doesn't filter his wines, because he considers that too "interventionist." Essentially, his evolution as a winemaker dovetailed with the worldwide rise of natural-wine techniques popular among the younger wine professionals. Meanwhile, members of the old guard, like Robert Parker, consistently attack or mock these popular natural wines.

In 2008, Tschida experimented with the ancient technique of aging in clay amphora called *qvevri* acquired from the former Soviet republic of Georgia, where they claim winemaking goes back 6,000 years. These amphora are thought to be the original aging vessel for wine. Yet while amphora use is another growing worldwide trend, he abandoned the practice after only a few years. "I didn't like the idea of buying these amphora from poor Georgian people and taking their culture," he said. "You shouldn't be making wine like this unless it's part of your culture. I was not happy, so I sold off all of my amphora."

Now, Tschida was exploring how little intervention he can bring to the winemaking process. "I would like to remove myself entirely from the winemaking. This is my next step. For a real natural wine, the human would have to accept anything that happened. In human history, there's really never been a truly natural wine." To that end, Tschida told me that he stopped tasting some of his wines before they are blended and bottled. "Even my taste is an intervention," he said.

As he said this, Tschida perhaps could see the skepticism wash over my face, my amused smile and eyebrow cocked. I could almost feel an eye-roll coming on. He laughed and said, "Look, in the end, I do it the way I do it."

As he opened another bottle, I asked Tschida if I could use his bathroom. "The toilet is down the hallway and you can use that," he said. "But I would be honored if you pissed in my garden instead."

"Wait, what?" Again, I searched his face to see if he was putting me on. "Really?"

"Yes, go ahead. I would be honored."

Biodynamic winegrowers are known to farm by the moon cycle and bury goat horns full of manure in their vineyards—but I had no idea what this was all about. In any case, it was now dark, so I walked outside into his dimly lit garden. Across the courtyard, I could see into the cellar, the large barrels holding Tschida's bleeding-edge wines. I turned my back to kitchen door, unzipped my pants, and starting peeing under one of his trees. I'd been drinking all day long, my bladder was full, and now the wine was now literally passing through and out of my body.

Perhaps it was because I was using his garden as a urinal, but I started to think of Tschida as a sort of Dadaist. He did go to art school, after all. Maybe his wine was simply part of a larger conceptual art project?

When I returned to Tschida's kitchen from the courtyard, Tschida had opened and poured a wine labeled Non-Tradition, which happened to be made with grüner veltliner. By now, I felt like I knew grüner veltliner pretty well. But nothing I knew about the grape prepared me for this wine. The word *unfiltered* was on the label. I swirled the wine, amber in color and very cloudy. It smelled funky and smoky. In contrast, the taste was subtle and airy, but hard to pin down. Once again, it was more a feeling than a flavor, something oily and nettly and complex. It felt ancient and mysterious on the middle of my tongue.

I tried to express this to Tschida. "Yes, the middle palate is always the truth of the wine," he said. "On the front of the mouth and the cheeks you can make it artificial. But the truth is always in the middle."

"Do you have to put *unfiltered* on the label?" I asked.

"No, I have it there so people can't complain afterward that they have sediment in the glass," he said with a laugh. "But it's also a mark of quality for me. This is a message wine. So many people in Austria talk about tradition, and haters say, 'This is not grüner veltliner.' I don't like to say it, but this is a message wine."

After we finished tasting Non-Tradition, we walked across the street from Tschida's house to have dinner at a country inn. We sat outside and looked across the vineyards at Lake Neusiedl and ate fried chicken and potato salad and drank his blaufränkisch, that godforsaken grape. Good blaufränkisch often reminds me of nebbiolo, but this was a different feeling, as if we were tasting a grape from further back in time. In the dark, we could see the outline of an old stone well, and beyond that reeds softly undulating in the breeze. "You know, as this century rolls on, with all our technological changes, wine like this will be the only real thing that still exists," Tschida said. "I really believe this. Wine won't change. The real thing will still have to exist, just like in ancient times."

I had to catch a taxi to the train station so I could return that evening to Vienna, but the wine and the vibe were so pleasant, we ended up finishing the bottle.

As we settled up, Tschida admitted that most of his fellow Austrians don't understand his wines. In fact, he sells much of his wine in Scandinavia and Japan. "Japanese people see my name and think I'm Japanese," he said. Then, he told me a crazy story about a recent trip to Tokyo. He got so drunk after a day of professional tastings and a long dinner, that when he returned to the hotel, he started running around the hallways, naked, growling like a monster and scaring the staff. He was asked to leave the hotel in the morning.

"Be sure to refer to me as enfant terrible," he said. "Some wine critic once wrote that and now it gets repeated in every article and blog."

Leon Trotsky found Vienna to be a most livable city. For the Marxist revolutionary and sworn enemy of the bourgeois, the stylish, imperial Habsburg capital was way more livable than, say, Czarist Saint Petersburg or some desultory city on the Black Sea or in Siberian exile. From 1907 until Austria-Hungary disastrously declared war on Serbia in 1914, Trotsky thrived in Vienna—"a city lavish with civilities"—living like a "lordly flaneur," sipping mochas, nibbling cakes, and smoking cigars in the city's ornate coffeehouses, where "repartee flashed from spoonful to spoonful of whipped cream," and where he scribbled notes about his conversations on Freudian psychoanalysis and scandalous artists such as Egon Schiele.

Trotsky's life in Vienna was one of "beautiful uselessness," to borrow the phrase that Joseph Stalin used to slander his rival. Stalin, in fact, spent a month here in 1913, holed up a few doors down from the emperor's summer palace at Schönbrunn. Stalin hated Vienna, as did another unhinged young man named Adolf Hitler, who in 1913 was painting and ranting in the city's Männerheim, the shabby "Home for Men," about nine tram stops past Sigmund Freud's apartment.

All this context is according to *Thunder at Twilight*, Frederic Morton's excellent book on Vienna at the precipice of World War I, in which he tells us that Trotsky penned despairing essays about his homeland. "How miserable was our gentry!" Trotsky wrote, worrying that his people back home were trapped by a dead-end "fanaticism of ideas, ruthless self-limitation and self-demarcation, distrust and suspicion and vigilant watching over their own purity."

More than a century later, I sat in Vienna's grand, stately Café Sperl (founded 1880) sipping a grüner veltliner on a rainy afternoon, reading Morton on my iPad. Around me buzzed stern waiters in uniform as people flipped through newspapers in several languages and played billiards, as if the Austro-Hungarian Empire still ruled Central Europe. As my trip coincided with the rise of Trumpism, I might have also noted that Trotsky's worries about his homeland mirrored my own worries about the intellectual and political direction of my own American people back home. But since this was not 1913, and I was not a revolutionary, I just ordered a rhubarb strudel and reflected on how much the city has changed since its fin de siècle heyday.

For instance, the formerly scandalous avant-garde art is not so scandalous anymore: posters for an exhibition of Egon

Schiele—once imprisoned for indecency and pornography—hung at tram stops. Blasé workers and schoolchildren started and ended their days amid images of a skinny, pale, nude woman displaying prominent pubic hair. For ten euros, you can visit the Sigmund Freud Museum and Instagram a jokey photo of the original cover of Dr. Freud's infamous treatise on cocaine, *Über Coca*. The traditional Viennese coffeehouse culture, during a period of *kaffeehaussterben* or "coffeehouse death" in the late 20th century, needed a little help from UNESCO to protect the city's "intangible cultural heritage"—ensuring these historic cafés survived into the 21st century, so foreigners like me can visit them to eat strudel and mispronounce fin de siècle.

What has not changed about contemporary Vienna is its livability. The city almost always tops those annual "world's most livable city" lists. I liked the mix of the grand imperial buildings and the quiet beauty of the Jugendstil architecture. I liked the clean streets, the amazing, punctual public transportation, and the lovely public spaces. I liked the people, who are politely aloof and standoffish at first, but who are lovely once they warm up to you. And because I am hopelessly nostalgic, I liked the melancholy history, which is inescapable. "Imperial Austria," writes Morton, "has become a byword for melodious decay." Most of all, I loved that it's so easy to eat well at low-key restaurants and drink great, relatively off-the-beaten path wines at local bars. I loved the casual dining of Vienna: the wood-paneled *beisl* with classic dishes like Wiener schnitzel, goulasch, spaetzle, and tafelspitz; the open-faced sandwiches at bustling Zum Schwarzen Kameel, which dates to the 17th century; even the "Wurstelbox," or sausage stands, for late-night currywurst and bratwurst.

And then there's the wine. Vienna, with about 1,700 acres of vineyards inside its city limits, is perhaps the only major capital city that produces wine. In fact, over half of the city's land is agricultural, especially in the outer districts north and west of the center, and particularly Stammersdorf in the 21st district and Maurer in the 23rd district. These cozy neighborhoods were once villages before being subsumed into Vienna proper, and they still keep a village feel.

Certainly the Habsburg monarchy loved its wines, brought to the capital from the far reaches of the empire—Royal Tokaji from Hungary, mountain wines from South Tyrol, aromatic whites from Friuli and Slovenia. But the true measure of the culture can be found at the *heuriger*, or traditional wine tavern—essential to the Viennese experience in a similar way as the Bavarian beer garden. *Heuriger* means "this year's"—as in "this year's wine"—and the concept dates back to an imperial decree in 1784, which allowed winemakers to open simple restaurants to sell their new wine. Many heurige, even now, are only open several weeks per year—you know they're open when the owner hangs evergreen branches outside. Morton, in *Thunder at Twilight*, describes the scene at a typical heuriger in 1913, which has changed only slightly since:

> *The proletariat of the outer districts, where their tenements adjoined the Vienna Woods, could stroll into green whose freshness ignored autumn. A few hellers would buy them a stick of horsemeat salami from stands by the vineyard inns; a few more coins would bring a bottle fermented from local grapes. Very soon they would lean against each other, sitting on the wine garden's hard wooden benches . . . enjoying an evanescent, convivial, inexpensive, very Viennese binge in a bower.*

After my Trotskyite flaneur fantasies over strudel at Café Sperl, I wanted to explore Austrian wine at its most elemental level. So one evening, I took a 20-minute tram ride west from the center out to the Maurer district. I knew I was at my stop when I saw an old grape press dating from 1800, next to a small produce market, and a big sign listing which heurige would be open that week.

In Maurer, I met my winemaker friends, Alex Zahel and his American wife, Hilary Merzbacher-Zahel, who took me on a quick tour of the Zahel family vineyards—which were literally sandwiched between apartment buildings and suburban dwellings, and which offered amazing views of the city center. "I like to bring people here," Alex said. "Sometimes when we talk about winemaking in Vienna, people don't believe how close to the city we are. Ninety percent of the people who visit think, 'OK, so you bring the grapes in from elsewhere and press them in Vienna?' But no. The vineyards are right here."

For our Maurer heuriger crawl, we began at the Zahel's own heuriger, seated at the *Stammtisch*, or regular's table, on wooden benches with pillows. We dined on pumpkin cream soup and roasted goose and pork schnitzel and drank Zahel's excellent gemischter satz, a white field blend of grapes like riesling and grüner veltliner, as well as more obscure ones such as rotgipfler, zierfandler, and neuburger. "A heuriger can be simple, but that's the goal of the heuriger," Alex said. "The Viennese mayor comes and shares a few glasses of wine at the same table as everyone else."

Afterward, we looked for the hanging evergreens, and ended up at Zahel's neighbor, Heuriger Lentz, where we met the proprietor, Reinhard Lentz. "In the past, we were only allowed to sell six dishes!" Lentz said. "Now, people don't even want to look at a menu. They know what they want." Beyond riesling, there were no noble grapes on the list. Besides the grüner veltliner and zweigelt and blaufränkisch, Lentz's menu included several grape varieties I'd never heard of, including blauburger and zweiburger. "That may be the only zweiburger in the world," Alex said. The thing is, these grape varieties weren't out of the ordinary here at all. What would seem weird and out of place to the patrons at Heuriger Lentz would be if Reinhard started serving Bordeaux blends or pinot noir or nebbiolo along with his sausages and schnitzel and sauerkraut.

Alex, Hilary, and I settled in with our glasses of zweiburger, blauburger, and müller-thurgau at a long wooden table, beneath deer antlers mounted on the walls. "Vienna is a small city. Maurer is a small village," Alex said. "You go to the heuriger to exchange stories." Reinhard, he added, is the village gossip. "Reinhard always knows everything. We had a small kitchen fire at our heuriger a little while back. It was minor. But before the fire department had even left, Reinhard was there, asking the firemen what was going on. He's the original citizen journalist!"

At our next stop, Weinbau Stadlmann, we met a woman in her late 50s named Frederika, who was a little tipsy on zweigelt rosé, and truly shocked that a stranger—someone not from Maurer, let alone from the United States, had stepped foot in her local heuriger. She introduced herself to me, and said, "It's an honor."

"No, no," I demurred. "It's an honor for me."

"Yes!" Frederika shouted. "It's not every day you meet a new person in a heuriger! When you don't live in a multicultural city, you can't wait to meet other humans!"

"Without my heuriger, I cannot live," Frederika said. "Not because of the wine, but because of the human beings I meet here. This is my living room. We make no dates. We just know that our friends and neighbors come to this heuriger. The owner has a heart like a deep diamond mine. He listens to my problems and I listen to his. This is our life! This is our lifestream!"

At Weinbau Stadlmann, there were still more uncommon grapes listed on the chalkboard: gelber muskateller, welschriesling, and sämling 88. We drank our wine with a guy named Wolfgang, a retired tour bus driver, who had clearly been enjoying a few glasses before we arrived. "Where are you from?" he asked.

"Near Philadelphia," I said.

"Fucking Philly!" Wolfgang said.

Wolfgang launched into a long, loud, tipsy saga of his own travels in the United States, back in the 1980s. He fondly recalled Seattle, San Francisco, and Los Angeles, partying poolside and on penthouse decks and riding in convertibles. As he held forth, Wolfgang told me that the highlight of his trip had been Hawaii, lying on the beach at Waikiki, and visiting Tom Selleck's house and "swimming in Tom Selleck's pool."

"It was the best two months of my life!" Wolfgang said. "I love America! I love Americans. They're so nice! People in Austria say, 'Ah, Americans are so bad.' But I say, 'Have you ever met an American? Have you been there?' I have. And it was the best two months of my life." Clearly, Stadlmann's other patrons had heard these stories many times before, but we were a new audience and glad to listen.

There's a well-worn word in German for this kind of cozy friendliness, good cheer, smiling warmth, and a sense of belonging: *Gemütlichkeit*. The ability to deliver this feeling, over a goulasch, a schnitzel—and most importantly a glass of wine—is perhaps Vienna's greatest virtue. In fact, it was noteworthy that everyone at heurige—men and women, young and old—drank wine. No beer, no cocktails, no whiskey. Few of the wines we drank at neighborhood wine taverns like Lentz or Stadlmann were Serious Wines, but that wasn't the point. The wines were drinkable, affable, and they generally cost two or three euros per glass. What the humble Austrian heuriger best represents is a culture that's at ease with wine. Unlike, say, ours back home in the United States.

Alex Zahel, in his mid-30s, has more ambition than to simply make wines for his neighborhood heuriger. So when I returned to Weingut Zahel a few days later, we delved deeper into their wines. We spent some time wandering the cellar, tasting straight from the barrel. Alex showed off some of his recent experiments with traminer grapes—gewürztraminer, as well as rarer varieties such as roter (red) traminer and gelber (yellow) traminer.

It was late afternoon, and the winery was bustling with loads of grapes coming from the vineyards. Alex was called away from the tasting room by his mother, shouting at him from the cellar below as the de-stemming machine made terrible noises and grapes were spilling over. Alex jumped on a forklift to fix the situation.

Upstairs in the tasting room, Hilary (who was once a magazine food editor in New York before marrying into a Viennese winemaking family) dove right away into the deeply obscure end of the wine pool. She opened a bottle of orangetraube, a dry, spicy, peachy white. Zahel grows what may be the only two hectares of this grape in the world. "This is probably the only 100 percent orangetraube in the world," she said. It's so rare that the Austrian wine law doesn't account for orangetraube as an official grape and so it can't be classified as a *qualitätswein*—just like Tschida's Himmel auf Erden, it's not permitted to bear the Austrian flag on its bottle top.

"What could warm the heart of a wine geek more than a grape so obscure that it isn't even included in the official classification?" said Alex, laughing as he returned from the cellar crisis. He's spot on. Wine geeks in the United States love Zahel's Orange T bottling. The Slanted Door, the famed Asian fusion restaurant in San Francisco's Ferry Building, has poured it for years. Wine blogger Patrick Ogle has described orangetraube as "a pinot gris that went over to a dry riesling's house in Vienna and had a threesome with a New Zealand sauvignon blanc." (OK, again, wine people are a horny bunch.)

"I didn't want to do this because it's trendy," Alex said. "I wanted to do this because these are the grapes that my grandfather planted." What exactly is orangetraube? "We don't really know," Alex said. "We believe it's in the traminer family; traminer is either the mother or the father. There was lots of crossing grapes in the early and mid–20th century. Winemakers were looking for high-yield wines. We know that goldburger, for instance, is a cross between orangetraube and welschriesling."

This is where ampelography begins to spiral into confusion and madness. Welschriesling means "Romanic" riesling (i.e., of the Romance-language countries of southern Europe). It goes by the aliases of riesling italico in Italy, graševina in Croatia, and olasz riesling in Hungary. However, welschriesling has absolutely no relation to riesling at all—its parents and origin are unknown. As for goldburger? The grape was created in 1922 by Dr. Fritz Zweigelt, a geneticist at the Klosterneuburg wine college who also created blauburger (a cross of blaufränkisch and blauer Portugieser) and rotburger (a cross of blaufränkisch and sankt laurent). Rotburger we now know as the doctor's namesake grape, zweigelt.

"They used to sell goldburger in two-liter bottles in the stores," Alex said. "My grandfather had a pillow near the dining room table and always hid a bottle of goldburger under it. He'd take a drink of it when my grandmother was in the kitchen. This was the time when people went out and drank two to three liters in an evening and then still drove home." A little bit of goldburger, along with neuburger, grüner veltliner, zierfandler, rotgipfler, and a lot of other grapes turn up in Zahel's most important wine: gemischter satz.

Gemischter satz, literally "mixed set," is the traditional white field blend of Vienna. All wine regions blend wines from various grapes, but this usually happens in the winery, long after harvest, just before bottling. What makes gemischter satz unique in the modern wine world is that the grapes are blended in the vineyard itself. Rows of three or ten or 20 grape varieties may grow in a single vineyard, and the grapes are all harvested, pressed, fermented, aged, and bottled together. Zahel's range of gemischter satz, for example, goes from a basic blend of grüner veltliner, riesling, and pinot blanc, to others with a blend of up to 20 varieties from five different vineyards.

For more than 200 years, gemischter satz have been poured as simple, crisp white wines in the heuriger. But in the past decade, a group of ambitious producers, Zahel included, pushed for an official classification and appellation, a *Districtus Austriae Controllatus* or DAC. They were successful, and 2013 was the first official vintage of Wiener Gemischter Satz DAC, the first time Austria had recognized a wine style as well as a geographic area. To be Wiener Gemischter Satz, a wine must have at least three and no more than 20 grape varieties in the field blend. The first grape must constitute no more than 50 percent of the blend, and the third grape must constitute at least 10 percent of the blend. These are meant to be bright, lively white wines aged in stainless steel; no Wiener Gemischter Satz can be more than 12.5 percent alcohol, and it must not have a "strongly recognizable expression" of oak. More than a quarter of Vienna's vineyards are now planted for gemischter satz.

I'll admit that I've fallen in love with both the egalitarian ideal of gemischter satz, as well as how consistent the wines are from top Vienna producers. These were honest wines, balanced with crispness, flavor, and a lacy lightness. The day after my evening with the Zahels in Maurer, I paid an afternoon visit to Rainer Christ, another producer who was instrumental in gemischter satz's rising reputation.

Rainer Christ, like Zahel, is a thirtysomething winemaker who runs a family heuriger—this one in Floridsdorf in the 21st district to the east of the city center. It's tucked on a leafy street, but right around the corner from a typical urban milieu: a busy bus stop, Show Time hairdresser shop, a horsemeat butcher, Kebab König, and a public WC. Inside, Christ's heuriger is modern, with white walls and a sleek wine bar adjoining it.

But it wasn't always so posh. Rainer told me when his family opened its heuriger back in 1927, it consisted of one table in the kitchen and one table on the veranda. They were open only 14 days in spring and 14 days in the fall. It wasn't until the 1950s that the family built a proper tavern. "That was mainly so the guests didn't have to use the family's private bathroom anymore," Rainer said.

Now, at 4:30 in the afternoon, Heuriger Christ was filled with big tables of middle-aged men and women, drinking and talking loudly. At a wooden table, we talked over glasses of Christ's Bisamberg Wiener Gemischter Satz 2014. The blend included grüner veltliner, weissburgunder (otherwise known as pinot blanc), riesling, welschriesling, "and a homeopathic dosage of several other grapes," he said, with a wink. It was a beautifully crisp wine, but with deep, rich, exotic flavors like smoked pineapple. "The character of the grape varieties steps back, and the character of the geographic area takes over," Christ said, adding that this wine came from 70-year-old vines in his prized Bisamberg vineyard, high above Vienna.

"We believe this is the roots of winemaking, and we want to bring it back to the market. Since 2006, I've been replanting all my vineyards with field blends," Christ said. "Gemischter satz is a forgotten style. After World War II, everyone was ripping out these blends. It became a monoculture. But 100 years ago, there were only field blends. These were the most famous wines of the monarchy."

Since all the various grapes are harvested together in gemischter satz, there will always be varying levels of ripeness each vintage. The advantage, Christ said, is that the differences between vintages should not be so dramatic. "The vineyards regulate themselves. The decision comes from nature. It's not crafted. It's not designed." For those who are concerned with biodiversity, the gemischter satz field blend is a way for lesser-known grape varieties to coexist with more popular ones.

As the heuriger grew more and more crowded, Christ's phone rang, and he was needed in the winery. He apologized, before leaving me with this thought: "Gemischter satz is a wine for people who rethink their drinking method. Maybe they'll realize they've been listening to a single instrument, and they realize they want to hear the whole orchestra."

A few days later, on a chilly Sunday morning, Alex and Hilary Zahel drove me to Gumpoldskirchen, a pretty town dating to 1140, near the Vienna Woods about a half hour south of the city. Many of the Emperor's prized wines came from the vineyards around Gumpoldskirchen, whose name was still well known by wine connoisseurs in the mid–20th century. But the region was hit hard by the 1985 wine scandal, and its sweet blends of local grapes like zierfandler and rotgipfler fell out of favor and were then forgotten. For a good deal of the late 20th century, many growers replanted chardonnay.

In its heyday, this area was known as the Südbahn, but it has now been vividly renamed Thermenregion, a nod to the thermal springs of nearby spa towns like Baden and Bad Vöslau. Gumpoldskirchen is just down the road from Mayerling, the scene of a tragic episode in Austrian history. There, the Crown Prince Rudolf, son of Emperor Franz Joseph, killed himself and his teenage mistress at the royal hunting lodge in 1889. Prince Rudolf's death devastated the imperial family, and many believe it set events in motion—including Archduke Franz Ferdinand becoming heir to the Habsburg throne—that ultimately led to World War I and the fall of the empire.

Arriving in Gumpoldskirchen, we pulled up to a half-timbered building, dark wood atop white walls, that dated to 1905. Here, at Weingut Spaetrot-Gebeshuber, we met Johannes Gebeshuber, a winemaker trying to revive interest in zierfandler and rotgipfler, the two local grapes blended into the famed wines of a century ago. Gebeshuber, now in his 40s, had acquired this building 20 years ago from a failing cooperative. In the early 20th century, the winery produced three million bottles per year. It was the official wine cellar for the city of Vienna, controlled by mayor Karl Lueger, a populist and anti-Semite who historians say was an inspiration for a young Adolf Hitler. But Lueger also ran restaurants in the city, and Gebeshuber says, "Maybe he was double-dipping a little too."

Gebeshuber took us on a tour of the cavernous cellar, amid beautiful, huge dark-wood casks engraved with detailed images of Saint George slaying the dragon (from 1907) and the Last Supper (dating to 1937). "When I started here, I grew 25 varieties. But I decided to focus only on zierfandler and rotgipfler," he said. Perhaps stating the obvious, he added: "We had hard times when we started with the local grapes. In years past, people were asking for chardonnay. People said, 'We don't know zierfandler and rotgipfler.' You couldn't sell it."

Alex Zahel said he could remember, back when he was in school, visiting his older colleague when Gebeshuber was struggling to find interest in his wines. Zahel said, "I looked at you and thought, 'Why did I ever want to be a winemaker?'"

After the cellar tour, we walked down the street and through a leafy courtyard to the Gebeshuber heuriger, which is run by Johannes's ex-wife, Johanna. We took a table amid the busy Sunday lunch buzz. The Zahels commented on how modern the interior design was for a heuriger, with its sleek white walls. "How old is this place?" asked Hilary Zahel.

"Oh this place?" Gebeshuber said. "Only about 400 or 500 years old." He opened a half-dozen bottles, we ordered a lunch of pork and chanterelle mushrooms, stuffed cabbage, and knödel filled with speck. And then we got down to tasting.

Zierfandler and rotgipfler are both strange grapes that make white wines. Rotgipfler is so named because the tip of its vine shoot is red. Zierfandler's local name—*spaetrot*, or "late red," from which Gebeshuber takes his own winery's name—is so called for the telltale red color of its skin when fully ripe. Zierfandler was grown all over the Austro-Hungarian Empire, where it was also called by its Hungarian name, *tzinifándli*. Some believe that what Americans call zinfandel, which is of Croatian origin, takes its name because that black Croatian grape was mistaken as *tzinifándli*, then wildly mislabeled as "Black Zinfardel [sic] of Hungary." When the Habsburg Imperial nursery first shipped "Black Zinfardel" to a New York nursery in the 1820s, the mistaken name spread.

In Gumpoldskirchen, rotgipfler, with its ripe fruit and unctuous body, has always been balanced and blended with zierfandler, with its acidity and minerality. "Years ago, these varieties were harvested very late," Gebeshuber said. "So the wines were too sweet and heavy with high alcohol." The wines he poured on that day were the opposite, completely dry and bright. These were happy wines. The younger blends were slender and carefree, while the ones with some aging on lees or in oak were chubbier, but graceful and generous, with notes of exotic tropical fruit.

The sun came out, and the combination of the classic Austrian cooking and the wonderful wines put everyone in a joking mood. We ogled the day-trippers from Vienna streaming in, wearing hunting jackets, hats, and big boots. "You always know the people from Vienna," Gebeshuber said. "They always come to Gumpoldskirchen dressed like farmers or hunters."

On that day, it was hard to believe very few people outside of Austria knew about these wonderful wines made from zierfandler and rotgipfler. And that thought made me once again feel a twinge of nostalgia for the Habsburg Empire's "melodious decay." I tried to imagine a world in which Gumpoldskirchen wines would still be sought after. It was nearly impossible. It would have had to involved, I guess, the alternative history of Austria-Hungary not being decimated during World War I. Surely it became much harder to make wines in Gumpoldskirchen after 1918, and even more difficult after the Nazis annexed Austria in 1938. Then came the vicious fighting between German panzers and the Russians during World War II, the postwar Allied occupation until 1955, and then living in the shadow of the Soviet satellite states just a few miles away. As the 20th century wore on here, one can almost—almost—empathize with the desperation that could have driven people to add diethylene glycol to sweeten their wines.

My mind came back to the heuriger. Alex Zahel had hopped on a tiny tricycle and was pedaling around the courtyard as we laughed and finished our wines. I asked Gebeshuber why he thought the Gumpoldskirchen wines had become so obscure.

"Fifty years ago, the producers here in Gumpoldskirchen didn't try their best," he told me. "That was the problem. They thought the glory would be forever."

CHAPTER EIGHT

The Meaning of Groo-Vee

WINE HAS ALWAYS BEEN GOVERNED BY FASHION—just like lapel widths, hemlines, and shades of lipstick. To grasp why so many grapes remain obscure or underappreciated, it's essential to understand how certain wines have fallen in and out of fashion. Consider, then, the 20-year trajectory of grüner veltliner.

Nearly 20 years ago, grüner veltliner was the new black. Then about ten years ago, it was de rigueur to dismiss Austria's premier white wine grape as a passing fad. Now (at least as this book was being written) it's fashionable in the wine world to once again profess admiration for grüner veltliner. The fickle in-crowd of sommeliers and wine writers have now "reconsidered" the variety and settled on the term "classic" to describe grüner veltliner—as if it were a Brooks Brothers navy blazer or Chuck Taylor All-Stars.

This makes me smile and roll my eyes at the same time. I want to be clear about one thing: I am not fickle. I'm being sincere when I tell you this: I have always loved GV. Always. I'm not one of those wine writers who fell quickly in and out of lust, only to newly "revisit" the grape lately because I needed a new story angle. I am true of heart. Please believe me, GV. I have never, ever stopped loving you! I love how diverse, how unpredictable, your mood and character can be, sometimes rich and aromatic, sometimes crisp and mineral, often peppery or spicy, and always lively, with big flavor without being too fruity. I love how friendly you are with food, how you step back just a little and let the food do the talking, and how you pair with just about everything, from fried chicken to barbecue to guacamole to sushi to pad Thai to tandoori. You even play nice with salads and vegetables, like asparagus, which notoriously don't get along with wine. Even chardonnay lovers might consider straying if they tasted you.

I look back fondly to the late 1990s, when grüner veltliner was just becoming trendy. I was still a young man, but had passed through my grunge-era, flannel-shirt-wearing, failed-novelist days and had begun a semirespectable career as a food writer. Grüner veltliner dominated the wine lists of the restaurants I was reviewing. "If viognier and sauvignon blanc had a baby," we were told, "it would be grüner veltliner." In many people's minds, it replaced both the New Zealand sauvignon blancs that were so popular and California viogniers that many were pushing.

Back then, GV seemed to appear out of nowhere, the archetypal "stranger comes to town" story. Prior to the 1980s, one of the only references to the grape that I can find comes in a 1978 column by the *New York Times*'s wine critic Frank J. Prial, who spells the grüner veltliner grape as "gruner weltliner" and dismisses it as a "fresh, light wine of no particular character." A few years later came Austria's poisoned wine scandal, and by the early 1990s, as Courtney Love, trip-hop, and Kate Moss trended, no one talked at all about Austrian wine, let alone an obscure umlauted variety.

Then, near the end of the Clinton administration, GV just exploded in popularity. In January 2000, Prial reported from a "big professional tasting at the Tribeca Grill," along with sommeliers "mostly in their 30s, looking like a gaggle of graduate students with their book bags and parkas stacked haphazardly by the door." Those sommeliers were after grüner veltliner, which Prial declared "was the most sought after wine at the tasting" and that "the once-inoffensive little wine of the Viennese cafés" was now "the rage" at New York restaurants. That rage burned on for the next several years. People started referring to grüner veltliner by the nickname "Groo-Vee" or by its first name only, "Grüner"—like Prince or Cher. Or perhaps more like Björk. "Grüner" actually just means "green" in German, referring to the *green* variety of the *veltliner* family of grapes.

Yet as the aughts and the Bush administration wore on, the love affair with GV waned. Cheaper and cheaper grüner veltliner started appearing on shelves, much of it in oversize, liter bottles. In 2006, wine writer Lettie Teague was asking, in *Food & Wine* magazine, "Is Grüner a Great Wine or a Groaner?" Teague quoted a top sommelier saying that grüner veltliner had become "too trendy" and "was kind of a one-night stand for me." By 2009, the bloom was officially off the rose when *Times* critic Eric Asimov wrote about a "disquieting" tasting in which his panel found "too many wines that were not up to snuff . . . some were ponderous and heavy . . . Others seemed simply wan and lacked snap." By 2010, grüner veltliner had ceased to be cool. The new wave of hipster sommeliers discovered Friuli orange wines or Jura *vin jaune* or rediscovered Mosel riesling or Loire chenin blanc or sherry. Or they moved on to sake or mezcal or bespoke cocktails or artisanal cider. Or in any case, on to other drinks trends—some of which will also eventually face the vicious cycle of becoming overhyped, losing their cool, and facing an inevitable backlash.

A few years later, in 2013, I read a fascinating piece of wine writing in an unlikely place: the importer's catalog of Terry Theise Estate Selections, amid tasting notes and wholesale prices of Austrian wines. In his catalog copy, the 60-something Theise addressed a new generation of wine professionals, millennials in their 20s and 30s, whom he saw ignoring and disrespecting grüner veltliner. About GV, he wrote:

> *"Most of you know it exists, yet there's a kind of stink to it, as in something that 'used to be trendy.' Think of the way you're discovering all these hitherto-unknown cool things from all over the place, and how much fun it is. That was Grüner Veltliner in the late 90s and early 'aughts.' And you don't want to repeat what those guys did; you want to do new things. Got it, and sympathize.*
>
> *The problem is, what should have happened was to recognize GV as a classic, whereas what did (too often) happen was it got swept into the rubbish pile of the previously fashionable.*
>
> *You're not gonna like what I'm about to say, but in the service of truth I have to say it. Not one single thing that's been discovered, trumped, lionized, promulgated, put on wine lists and talked about with giddy delight, not ONE. DAMN. THING. has been nearly as excellent as Grüner Veltliner."*

Theise, as it happens, was one of the key importers of Austrian wine, and the man who introduced many of us to grüner veltliner in the 1990s. Jay McInerney—It-novelist-turned-wine-writer—once referred to Theise, in his *Wall Street Journal* column, as "near the pinnacle" of wine "hipness." I found Theise's plea to a younger generation of wine buyers noteworthy, possibly even poignant. I could so clearly picture this middle-aged man, frustrated that he can't get the kids to see how cool it was, way back when. Theise, who created the fashion for Groo-Vee in the first place, was now caught in the vicious trend cycle that rewards the flavor of the month and punishes the flavor of last month.

But a curious thing happened. Not too long after Theise's plaintive catalog copy, a new sort of reconsideration of grüner veltliner seemed to take hold. "It used to be that I considered Grüner Veltliner a fad grape. I've come to see just how much it can make a world-class wine," said Jon Bonné, then wine editor of the *San Francisco Chronicle*, in an interview, exemplifying this critical stance.

Around the same time, the Austrian Wine Marketing Board sponsored a fancy trade tasting of older grüner veltliner, with vintages from the 1970s through 1990s, at Le Bernardin in New York. The open purpose of this was to refurbish GV's reputation. Many of wine's gatekeepers and heavy-hitting wine writers were in attendance—including Theise, whose wines were, of course, being poured. Apparently, the wine people in attendance reconsidered grüner veltliner and came away raving. (I'm getting this second hand; I was not heavy-hitting enough to be invited). After swirling, sipping, and spitting, then presumably noshing on a few hors d'oeuvres, the gatekeepers and critics dutifully went home and wrote glowing things about grüner veltliner's "ageability" and "versatility" and "value" in articles and blogs and social media. Suddenly, the wine world was chattering again about GV, old Groo-Vee. Except Groo-Vee didn't go by nicknames anymore, just Grüner Veltliner, thank you. It was sort of like when an old college friend—whom you once called Jimmy or Billy or Bobby—gets a good job after graduation, starts wearing a suit, and now requests to be called James or William or Robert.

Jancis Robinson, who was at Le Bernardin tasting grüner veltliner that day, was amused enough to write a column for London's *Financial Times* titled "New York's faddish wine community." "I was in Manhattan two weeks ago and was fascinated to observe just how fashion-conscious its wine commentators are," Robinson wrote. "Image is everything in the faddish New York market. If Groo-vee had been less popular it might have chugged along as a welcome ingredient on any wine list, as it is in the UK, but success can be a killer in New York."

As I was writing this book in the spring of 2017, Jon Bonné reported in the trends-obsessed drinks blog *Punch*, that it was "undeniably, a moment to reconsider grüner veltliner. It is, perhaps, the most important white wine that people don't know they want to drink." Of the six wines he recommended, four were imported by Terry Theise.

All of which points to a lesson we probably already know: When it comes to culture, what is currently in fashion will soon fall out of fashion; once it's out of fashion, it will likely come back around into fashion again. Just wait long enough, like I have with my old grunge-era flannel shirts.

I wore those flannel shirts during the two weeks I spent in Lower Austria during the chilly autumn of 2015. I had been following my passion for arcane and underappreciated grapes for several years at this point, but Austria kept calling me back. Austria felt essential to understanding what wine had been in centuries past, as well as where modern wine might be headed. I wanted to experience the obscure grapes that I loved, in a place where they weren't actually obscure.

On my trip, I visited nearly 40 winemakers, traversing wine country in a rented Skoda. Over the first few days, I noticed an odd phenomenon: Whenever I was at a stoplight, or circling a roundabout, or entering a highway, I received a good deal of hostile honking from fellow drivers. At one point, an old man shook his fist at me as he passed me on the outskirts of Vienna. I became self-conscious and hyperaware of my driving, worried that I was unwittingly violating some unwritten Austrian automobile code.

The mystery was finally cleared up one morning when I arrived to meet a winemaker at his cellar. When the winemaker approached my car to greet me, he seemed confused, and asked, "Did you come from Slovakia?"

"No," I said. "What makes you ask that?"

"Hmm," he said. "Did you know that your car has Slovakian license plates?"

I hadn't noticed, but now that I looked at the plates, I could see he was correct. With a chuckle, I told him about how I'd strangely and consistently been honked at throughout my trip.

"Yes, well you have Slovakian plates," he said, as if this explained everything.

Austria was tense that fall. For several months, desperate refugees from Syria had been pouring into the country via the Slovakian, Hungarian, and Slovenian borders. The far-right, anti-immigrant Freedom Party was steadily gaining support in advance of elections.

Things were tense in the vineyards as well. The prior year, 2014, had been a bad one for winegrowing along the Danube. The official 2014 vintage report had been cagey, saying "nasty conditions weren't in short supply" and talking about "challenging" heat, "difficult" cloudiness, and "problematic" rains, and only hinting at devastating hailstorms and a monsoon-like September. But the report depicted a dramatic, heroic "struggle" that still managed to produce "pleasurable" or "lean" or "austere" wines, still assuredly delivering "value" and "quality." In reality, 2014 was an unmitigated disaster for many Austrian winemakers, particularly for red wine. In the blunt words of one red-wine producer I visited: "2014 was shit. I can't remember such a vintage." Even many white-wine producers lost more than half their normal grape production.

And yet, a year later, as they started to bottle white wine in many cellars, a cautious murmur arose: The 2014 grüner veltliners actually tasted pretty good. It was a small harvest, a fraction of the usual volume, but there was quiet hope that these wines might age as well as any in recent memory. Georg Grohs, who works at the Viennese winemaker Wieninger, used a sailing metaphor to describe the vintage: "When there's wind, every idiot can sail. When there's no wind, only the best can sail."

Suffice to say that I tasted a boatload of 2014 grüner veltliners as I navigated my Slovakian Skoda from Vienna along the Danube, through the prime grüner veltliner lands. But the vibe was decidedly different from the Groo-Vee of ten years before. "There's a big discussion right now about what grüner veltliner should be. Should it be a serious wine?" Martin Moser, winemaker at Hermann Moser estate in the Kremstal, told me. As I moved from Kremstal to Kamptal, from Weinviertel to Wagram to Wachau, what struck me most was an utter seriousness of purpose. By and large, the better producers of grüner veltliner seemed determined to evolve the grape far, far away from the "Groo-Vee" vibe and to climb the ladder toward becoming a truly Serious Wine.

On the overcast, windy day that United States Secretary of State John Kerry visited Vienna to discuss the immigration crisis, I went to Wagram to visit Anton Bauer on the last day of his grüner veltliner harvest. As we strolled his vineyards, leaves turning yellow, Bauer pointed out huge black nets, costing thousands of euros, that he'd draped over his vines to fend off devastating hail, which cost his neighbors over half their grapes last year. He told me a story about sommeliers at London's Capital Hotel who'd blind-tasted his reserve grüner veltliner and confused it with grand cru Burgundy.

Bauer also had me taste his fuller-bodied roter veltliner, or "red" veltliner, which grows on a few hundred acres in Wagram. There are also rarer veltliners, frühroter ("early red") veltliner and the nearly extinct brauner ("brown") veltliner that you also occasionally find in Austria. Grüner veltliner, despite its name, is actually not related to any of them—thus, once again highlighting the lunacy of grape nomenclature.

In fact, the origin of grüner veltliner remains a mystery. It's a half-sibling to rotgipfler, the historic grape of Gumpoldskirchen, with which it shares savagnin as a parent (savagnin being the "godforsaken" grape also known as heida or traminer). But grüner veltliner's other parent has only recently been discovered through DNA testing of a 500-year-old vine unearthed in a Burgenland village called St. Georgener. This St. Georgener Vine, dating to at least the 16th century, was found in 2000 by a man named Michael Leberl, who'd heard stories as a child of an ancient, aromatic vine from his mother and eventually located it as an adult with the help of village elders. Some believe the vine is the legendary *grünmuskateller*, or green muscat, that disappeared long ago. The St. Georgener Vine is so robust that it actually survived being hacked to pieces by vandals in 2011. Within months, its stems sprouted new shoots. A few years later, 300 liters of wine were produced from its grapes, and it's now protected as a national monument.

Later, as my Skoda and I got lost, I ended up over an hour late to visit Josef Ehmoser and his wife, Martina. Josef and Martina, gentle and quiet, had patiently and kindly awaited me. Ehmoser told me that when he first returned from school to take over the family vineyards, he'd planted sauvignon blanc and chardonnay. But after a few years, he decided to rip those international grapes out and focus instead on grüner veltliner and riesling. The Ehmoser wines were also quiet and gentle at first, but they had a sharp backbone of precise, linear acidity, with lively spice and white pepper on the finish. "We want to have wines that are not everyone's darling," Martina said. "It's important for us not to have easy-drinking wines."

"This is a grüner veltliner to think about," said Martina, as she poured their single-vineyard Georgenberg. I took a long time tasting that powerful, intense wine. Josef and Martina sat in silence. Finally I said, "If I tell you that your wines are very serious, I hope you take that as a compliment."

"Oh yes," they said, furtively looking at one another and smiling shyly. "Yes, serious is a very fine compliment."

A few days later, on a sunny late October day in Weinviertel, northeast of Vienna, I visited with Marion Ebner, a striking and stylish winemaker who runs Ebner-Ebenauer with her husband, Manfred. The impeccably designed Ebner-Ebenauer home, where we tasted ambitious grüner veltliner at a thick wooden table, sitting on transparent chairs, with Ella Fitzgerald on the stereo, and Ebner's perfect dog Moka lying underneath, could have been featured in *Metropolitan Home*. So precise was everything that above the toilet, there was a pictorial sign suggesting that men sit down to pee. Ebner talked about her wines like children. One, from the Sauberg vineyard, she called her "problem child." She said, "Sometimes he's a bad boy and he's so stinky. So we know when he's stinky, he needs to stay in the barrel a little longer."

I was surprised to hear that Ebner was experimenting so much with aging grüner veltliner in barriques, or small barrels that can impart more aggressive oak character to the wine—something many Austrian producers, who age GV in stainless steel or larger casks, scoff at. She shrugged and boasted that the barrique wines were among the highest scores she'd received from Robert Parker. "At first, people laughed and said, 'Look at this girl who makes a grüner in a barrique.' And then I got the Parker points, and they don't laugh anymore."

The following week, in Kamptal, I visited the tasting room at Hirsch, with its floor-to-ceiling plate-glass windows opening up to epic views of Heiligenstein and Lamm, two of Austria's most prestigious vineyards, where both grüner veltliner and riesling grow. In 2010, both Heiligenstein and Lamm were among the first vineyards in Austria classified as *Erste Lage* or "First Growth," signifying the country's finest terroir. This is similar to Bordeaux's Premier Cru or Burgundy's Grand Cru or Barolo's crus. Classifying *cru* seems to be an obvious first step that any wine region takes when it moves toward a Serious Wine. If grüner veltliner is meant to be taken as seriously as nebbiolo or pinot noir, the gatekeepers and collectors need to be assured that the grape shows off myriad expressions of terroir—which it does. Serious Wines must always offer new knowledge, a rabbit hole of villages and vineyard and vintages for experts and aficionados to slip down.

I tasted with Johannes Hirsch as the sun set over the Erste Lage vineyards, which he farms using strict biodynamic methods. A fortysomething like me, Hirsch has been on the cutting edge since the original rise of GV. For instance, he was one of the first to use screw caps to bottle his best single-vineyard wines, which was controversial at the time, but is now widespread among Austrian winemakers. Hirsch now showed me his new, attractively minimalist labels. Until recently, Hirsch's labels on his entry-level wines featured whimsical cartoon deer (Hirsch means *deer* in German). This was the Austrian version of the funny "critter label" trend begun by mass-market producers such as Yellow Tail. But not anymore. "People see a fun label, and they think it's a supermarket wine," Hirsch said. "I don't want to say this isn't a fun wine. But there is serious wine in this bottle." Ironically, in the 2006 *Food & Wine* article that asked "Is Grüner a Great Wine or a Groaner?" Hirsch was quoted while showing off the cartoon deer label, which he'd only recently introduced. "When we changed the label," he told *Food & Wine* in 2006, "we sold five times as much wine." What a difference a decade makes. Perhaps it's not surprising to learn that Hirsch is imported into the United States by Terry Theise.

None of those fashion decisions seem to affect the juice in Hirsch's bottles. His grüner veltliners—which he gently ages in large oak casks as well as stainless steel—were wonderful in 2006, and over a decade later they still are amazing. Hirsch wines begin in dulcet, precise, soothing tones, but they finish in a crescendo, full-bodied, deep, and powerful. It's as if they flex their muscles around the midpalate. One of my favorites was from Grub, another Erste Lage vineyard: floral and fresh, with a hint of smoke on the nose, but then finishing fleshy, ripe, and bone dry. It lingered like some distant memory of an ideal summer night, peaches grilled over a campfire in a wildflower field. "Wow, this a sexy wine," I said.

"Well," Hirsch said. "You know that the wines always take on the character of the winemaker." After we had a laugh over that cheesy joke, Hirsch grew contemplative, and considered the challenges that grüner veltliner faces on the road to becoming a Serious Wine. "Look," he said. "Grüner veltliner is still a foodstuff to us. It's not yet considered a luxury product. We're still decades behind. Collectors still don't have the self-confidence to say, 'I like this. I'm buying this and not Burgundy.'" I believe that drinking a bottling like Hirsch's 2014 Grub in, say, 2024 or 2034 might finally begin to convince them. That, of course, is the conundrum: A Serious Wine is, among other things, wine that ages. But how will you ever know if it ages, unless it's taken seriously enough to age?

Of course, not everyone was convinced that the new seriousness was the right direction for grüner veltliner. "I believe making a good basic wine is more difficult than getting 100 points out of a special grand cru," said Rudi Pichler, a highly regarded producer in the Wachau. Of course, the Wachau—the gorgeous, picture-postcard, tourism-friendly, UNESCO-protected valley—is considered to be Austria's most prestigious appellation, its riesling and grüner veltliner world renowned. Wachau wines have been coveted since at least the fifth century. Roman emperor Marcus Aurelius Probus allowed wine to be made here, even though the Romans banned grape-growing most everywhere else north of the Alps. Some believe grüner veltliner vines arrived along the Danube during Roman times. In 1983, prior to the scandal, producers here had already created their own strict codex, the Vinea Wachau Nobilis Districtus, to govern the region's wine. As a result, the Wachau regained its footing much faster after 1985. So perhaps there's less urgency than in the up-and-coming regions.

At some of my visits along the Danube, the talk grew so serious that at certain points, I wasn't sure whether we were discussing wine or something deeper. One late afternoon, I tasted with another legendary Wachau producer, 40-year-old Emmerich Knoll. One of the greatest white wines I'd ever tasted was a 1990 Knoll grüner veltliner that I'd bought in 2014. If there was any question that grüner veltliner can live for decades, that 24-year-old Knoll wine put it to rest. Eighteen hours after I'd opened it, I brought what remained in the bottle to lunch with another winemaker who took one sip and exclaimed, "Ah, this is a Stradivarius!"

Knoll was a stern, no-bullshit guy. When I asked him, with a chuckle, "What do we talk about when we talk about grüner veltliner?" he looked at me like I was a fool. He contended that grüner veltliner's greatness lies in its diversity: You can't pin it down to one flavor or aroma, that its character and personality change depending on where it's grown. "You can taste ten grüner veltliners and still cannot say, 'I know what grüner veltliner tastes like.' It's the complete opposite of chardonnay." Above all, Knoll insisted that there was still too much cheap grüner veltliner being produced for the US market. "You can't be trendy with a thing like wine."

The sun began to set before five P.M. Our talk veered toward climate change, and the worry from critics and sommeliers that alcohol levels are rising in the Wachau, resulting in overripe wines that bring big, concentrated flavor, but are ultimately out of balance. Knoll dismissed these criticisms. "I think it's a pity that we take one single factor and say, 'We don't want this.'" He acknowledged that climate change presented challenges, especially in a wine region known for cool-climate wines. "But climate change does not simply mean it's getting hotter all the time. We get also get more rain than we did in the 1980s and 1990s."

As it grew dark, we began to be swarmed by the harmless fruit flies that appear during grape harvest time in Austria. "My biggest concern is simplification," he said, as the flies fluttered around us. "A lot of people believe in too-simple explanations for things. Starting with wine, and then moving on to political issues."

After leaving Knoll on that chilly night, I ate a quiet dinner alone in a modern *gasthaus* near the town of Dürnstein, not far from the ruins of a castle where the 12th-century English king Richard the Lionheart was imprisoned on his way home from the Crusades. I ate a bowl of rich pumpkin soup (sprinkled with pumpkin seeds, drizzled with pumpkin seed oil) followed by braised chicken in a creamy paprika sauce, a classic Hungarian dish that long ago found its way into Austrian cuisine, served with spaetzle. When I was finished and tried to settle up with a credit card, the waiter told me the restaurant only took cash and so I had to drive to the next village to find an ATM.

Finally, I returned to my hotel, which was lovely and surrounded by grapevines, but on this midweek night near the end of tourist season, silent and nearly empty. I was restless. I agreed with Knoll that the world in 2015 was growing in complexity and yet people were looking for increasingly simple explanations and solutions.

I decided to open a bottle of grüner veltliner I'd bought a few days earlier. My room opened up to a little patio, and I bundled up and took my glass and bottle outside. The wine was Ludwig Neumayer's *Der Wein vom Stein*, or "the wine from stone." It was difficult to put this wine into words, so pure that it was like tasting the rain or the wind. Purity is a word that gets thrown around a lot in the wine world, but Neumayer's, aged solely in stainless steel, has it—a depth and restraint that feels like old-time religion, like Johnny Cash singing "God's Gonna Cut You Down."

As I drank, I thought about the winemaker. Neumayer was reserved and soft-spoken, with gray-white hair, ruddy complexion, and square eyeglasses. He reminded me of a humble, old-school Protestant minister, the type you rarely find in America these days, the complete opposite of the loudmouth megachurch type. While we tasted, Neumayer tried to explain that he was shy, but momentarily blanked on the word in English. "What is the word?"

"Shy?" I asked. "You're shy?"

He blushed, bright red. "Yes, yes. I'm very shy. And I'm such a small producer. Some months, I don't even have enough wine to sell. I'm a bad marketing man."

I'd met Neumayer a few days earlier in the village of Inzersdorf ob der Traisen, in a relatively obscure wine region called Traisental. Traisental is considered to be on the "wrong" side of the Danube, down from Wachau through the hills and past the 11th-century Göttweig Abbey. While wine has been made here for centuries, it has only been an official Austrian appellation since 1996. Traisental has fewer Erste Lage than other regions like Wachau and Kamptal. Until about seven years ago, when a new bridge was built, it was still inconvenient to reach the area from Vienna.

"This is a very old place. We found a gold coin here from the 1400s," Neumayer told me as we roamed his oldest vineyard, Zwirch, with 60-year-old vines planted on a limestone hill 1,000 feet above the valley. In the middle of the vineyard sat a hunter's shack. "The deer are gourmets here," he said, with a wink. "They only eat the best grapes."

Neumayer began making wine in the dark days of the late 1980s. "1986 was actually one of the greatest vintages." Few outside of Austria would know this, of course. Years ago, he said, grüner veltliner used to be called weissgipfler in Traisental. *Weiss* (white) was to clearly distinguish its leaves from that of the red-tipped rotgipfler, its half-sibling and the more sought-after grape of Gumpoldskirchen, less than an hour down the road. "In the 1990s, grüner veltliner was just a good, normal wine. But then the hype came in the 2000s."

The bad 2014 harvest was even more difficult for Neumayer, who told me his mother lay dying in the hospital that autumn. "At the end of the harvest, my mother died," he said. "Of course, this is normal for the end of a long life, and she lived a good life." Neumayer told me she had been one of 18 children, and had lived her whole life on a farm.

"After the harvest, my niece asked if I could make a special wine in remembrance to my mother. And so I made this. There are only 1,000 bottles." He opened and poured it for me. "You've come all this way to Austria. So I have to pour this for you," he said, with a warm and sad smile. This wine, instead of Johnny Cash, was more like Tony Bennett.

"Would your mother have liked this wine?" I asked.

"Oh yes," he said. "My mother was a great wine taster."

Neumayer had always lived on the family farm with his mother. But he assured me that his life had not been a monastic one. "I like living here," he said, "but I also like to visit the big cities. I've spent a lot of time and money tasting wines from all over the world. Tasting is comparing. It's important to have the possibility of tasting and comparing."

"I'm fascinated with big cities like New York or Paris. It's always interesting to see how all the races and religions can live together without fighting." As happened so often in Austria that fall, the wine talk turned to immigration, and to the Syrian refugee crisis. "This is a new situation for us," Neumayer said. "But it's a fact of life that we have to deal with. We cannot turn away these people."

"But what do I know," he said with a sigh, followed by a wink. "I'm now told my wines are vegan wines, which is very modern and nice. But two years ago, I didn't even know what a vegan was."

CHAPTER NINE

Blue Frank and Dr. Zweigelt

"I DOUBT, THEREFORE I AM," said Roland Velich, winemaker at Moric winery. "It's good to doubt things or else there is no movement." Velich and I were tasting his blaufränkisch wines, but not in a typical tasting room. Rather, we sat in his mod, softly lit, blanc-and-noir study, essentially an intellectual's man cave. I sat on an expensive black leather couch while the *Goldberg Variations* played softly in the background, and Velich stroked the back of his beautiful nine-year-old caramel-colored dog, a Vizsla (or Hungarian pointer) that sat serenely at his feet on the white shag rug. As he poured each Moric blaufränkisch, each from a different vineyard in Mittelburgenland, each one of the finest expressions of the grape, he lined the bottles on his mantle: the intense Reserve, full of dark, brooding spice; the big, deeply earthy Lutzmannsburg, from 100-year-old vines, like picking berries in an ancient forest after a heavy rain; the decadent, ever-unfolding Neckenmarkt, from primary rock at over 1,000 feet altitude, supple and bright, with a long finish like delicate red fruits and cocoa sipped from a smooth bowl made of rare, exotic wood. Tasting these complex wines in Velich's man cave, I imagined myself a character in a sensuously philosophical European novel, say Milan Kundera's *The Unbearable Lightness of Being.*

While we tasted, we discussed old Robert Parker and his unhinged rant against "godforsaken grapes." Moric wines have always received very high scores from Parker's *Wine Advocate,* so it perplexed Velich that blaufränkisch had been singled out on the blacklist. He expressed skepticism over the certainty and entrenched opinions that shape so much discussion about wine from critics and sommeliers and collectors. "Doubting is the beginning of intellect," he said.

As the bottles lined up on the mantel, I certainly doubted Parker and his ranting assertion that blaufränkisch was a grape that "in hundreds of years of viticulture, wine consumption, etc." has "never gotten traction" because it has been "rarely of interest." This notion is simply false. The grape name blaufränkisch—literally "*blue* Frankish"—dates to the Middle Ages, when Charlemagne was King of the Franks and Holy Roman Emperor, ruling Europe in the eighth century from what is now Aachen, Germany. Fränkisch was a term of quality, differentiating it from things that were *Heunisch* ("from the Huns"), a pejorative describing anything from the eastern Slavic lands. Later, in the 12th century, Hildegard of Bingen, the German nun and mystic, wrote that wines such as blaufränkisch were stronger than Heunisch wines and had an effect on the motion of the blood. All of which means that blaufränkisch was considered a "noble grape" by a powerful monarchy much earlier than either pinot noir or cabernet sauvignon in Burgundy or Bordeaux. So either Robert Parker doesn't know his wine history (not very likely) or he's clinging to some 20th-century narrative about wine that isn't true anymore.

Velich's home and winery sit behind a modest church in the village of Großhöflein, in Mittelburgenland, 15 minutes from the Hungarian border. "This is where the German-speaking world ends, and the Slavic world starts," he said. "This is the end of the Alps and the beginning of the Carpathian mountains." I was reminded that Burgenland had been part of Hungary until 1922. "We are Hungarians here. This was once a great wine community. The most prized wines of the 19th century were Hungarian wines. But we are playing catch up. We lost 100 years." His family has made sweet wines on the opposite side of Lake Neusiedl for decades, but Velich left the family business years ago ("complicated politics") to make blaufränkisch in the higher-elevation, cooler-climate areas of Mittelburgenland west of the lake. "It's a very special place, but it's not as easy to make wine here as it is in, say, Spain"—Velich's grapes will always struggle to ripen, unlike those grown in the hot Iberian sun. He then added, echoing the sentiment of so many grüner veltliner producers I met: "The next big step, of course, is to prove that blaufränkisch can be a collectible wine."

While grüner veltliner has experienced a rollercoaster relationship with American wine drinkers, Austrian red wines have struggled to find love, remaining relatively obscure. Certainly, you'll see ageworthy blaufränkisch like Moric on great wine lists, and a good wine bar may offer an entry-level Austrian red by the glass. But there's rarely much buzz, and little understanding or interest outside the wine-geek bubble. Every time it appears that blaufränkisch, or fellow Austrian reds zweigelt and sankt laurent, are poised to step into the mainstream spotlight, something seems to hold them back.

For me, this is a shame. I believe that Austrian reds—cool climate, low alcohol levels, not too oaky, incredibly food-friendly—personify what it means to call a wine "drinkable." I'm baffled that more people have not embraced them, and I'm not the only one. "Sometimes, for one reason or another, an entire category of wine seems to be virtually ignored," Eric Asimov wrote, in the *New York Times*, in October 2015, right before I arrived in Austria, bemoaning the fact that too many people saw Austrian reds as "alien and suspect."

Sometimes I wonder if it's all a matter of naming. Blaufränkisch, with its umlaut and three syllables and foreign pronunciation, is always going to be a tough sell. In Germany they call the same grape lemberger and in Hungary they call it kékfrankos—neither much of an improvement. But after spending several days in Burgenland, a thought washed over me. Maybe what blaufränkisch needs is a nickname, just like Groo-Vee for grüner veltliner? Suddenly, the name Blue Frank popped into my head. Who wouldn't want to drink Blue Frank?

In Velich's man cave, I hesitantly floated the idea of giving blaufränkisch the nickname of Blue Frank. He looked at me like I might be a fool. Or at least definitely not a character in that sensuously philosophical European novel. Or if I was, I was the hapless American whose wife slept with the mysterious European. In any case, Velich smiled gently and said, "It's not about style. It's about substance. We're still trying to figure out what we are."

One of the very first mentions of Austrian red wines in the mainstream press came as recently as 2006, by the late bon vivant newsman R. W. Apple Jr., who declared in the *Times*: "Austrian reds must represent the height of obscurity." Things have changed only slightly since then. A major reason for their obscurity is obvious: Until about a decade ago, most reds from Austria were still not ready for prime time, over-oaked, overly fruity, overly concentrated. "Before 1990, forget everything that's red wine in Austria," said Franz Leth, a producer in the Wagram who makes excellent zweigelt. "1990 was, for me, the beginning of red winemaking here. At least by international standards." Even the boisterous importer Terry Theise, not too long ago, was not bullish on Austrian reds, striking a gentle note of skepticism. "Austrian red wine is to be taken seriously, that much is beyond dispute," he published as recently as his 2010 catalog. "Yet for every truly elegant grown-up wine there are many others that are silly, show-offy, insipid, even flawed." Velich was one of the producers who forged a more subtle path, less reliant on oak barreling and other cellar tricks. "I doubted that a great red wine from Austria needed to be made with lots of new oak," he said. "I doubted that we needed mechanization."

In October 2014, the Austrian Wine Marketing Board hosted another fancy tasting, this time to promote the red wines at The Modern, the Michelin-starred restaurant inside the Museum of Modern Art. At that point, I was apparently suitable for inviting, and I sat among the sommeliers and wine merchants and bloggers and scenesters as we tasted through 28 glasses of zweigelt and sankt laurent and blaufränkisch. The wines showed incredibly well, especially the last flight of five older blaufränkisch from Burgenland, dating back into the early 1990s. Some compare blaufränkisch to cru Beaujolais or northern Rhône syrah, but these older vintages bolstered my belief that good blaufränkisch can be the equal of nebbiolo. With their combination of juicy freshness, a savory core, surprising dark mineral notes like hot asphalt on a summer day, and beautiful aromas of rose petals, they could have easily fooled a lot of Barolo aficionados.

All around the private room at The Modern, I could see a lot of nodding heads and raised eyebrows. The vibe was decidedly different than most industry tastings, which can be deadly boring, often a room of middle-aged white guys, know-it-alls in wrinkled blazers mansplaining and asking nitpicky questions. But every once in a blue moon, a sponsored tasting can serve the same purpose as a retrospective show of an underappreciated artist, highlighting qualities that we've never noticed before, or even rewriting the art-history narrative. Perhaps it was fitting that we were at MoMA sipping and spitting great blaufränkisch.

I told Velich about this MoMA tasting, and suggested that blaufränkisch was perhaps like an underappreciated artist, a painter of overlooked masterpieces. He considered the idea with a wry smirk. "Well, art is art and wine is wine. Wine is not art. Wine is older than art."

After the MoMA tasting, during the final months of 2014, there was more-than-usual chatter about Austrian red wines. Yet even as the industry was ever so gradually introducing drinkers to blaufränkisch's potential as Serious Wine, the gulpability of zweigelt, or the cool conundrum of sankt laurent, the actual 2014 red wine vintage in Austria was turning out to be disastrous. I'd seen the disaster firsthand on a short trip to Austria in September 2014. I recall visiting a rotund, warmhearted winemaker named Walter Glatzer in a quaint village called Göttlesbrunn, in the red-wine region of Carnuntum, near the Slovakian border less than an hour south of Vienna. Walter drove me around his vineyards in a small truck, with an open package of sausages lying between us on the seat. It was a gray day, with a steady drizzle of rain, and Glatzer glumly surveyed his prized rows of zweigelt and blaufränkisch. I don't know that I've ever seen a man look so distraught. Glatzer ate his feelings, mindlessly munching the cold sausage while pointing out huge puddles of mud beneath his vines. "You can see where the big rains happened this week," he said. "We have to harvest right away, and even then it's too late. It's very sad. You work all year. Working, working. Then in the last ten days it rains so much and everything goes to shit." He said he'd probably lose almost 50 percent of the grapes in this vineyard. Later, back at the winery, we tasted an earlier vintage from that vineyard, which Glatzer called "his sweetheart wine." He told me, with a heavy sigh, "I'm 90 percent sure we don't make this wine this year."

All of the ups and downs of 2014 were firmly in mind as I navigated my Skoda south of Vienna through red-wine territory in the late autumn of 2015. Two weeks before Saint Martin's Day, I returned to Göttlesbrunn, rolling into town through narrow streets of pastel-painted homes, lined with pink and red flowers.

Saint Martin's Day, otherwise known as November 11, is an important day in agricultural regions across Europe. Since the Middle Ages, Saint Martin's Day has been celebrated as the end of the harvest season and the beginning of winter. Martin of Tours was a fourth-century Roman soldier from a town called Savaria—which is now modern-day Szombathely, Hungary, only about 15 miles from the Austrian border. Martin dropped out of the Roman army to become a Christian monk and eventually became the bishop of Tours, in France's Loire Valley. Some legends claim he introduced chenin blanc, which became the Loire's main white wine grape, to the region. Another myth tells of a winter night at a monastery: Martin's donkey got loose in the vineyard and ate all the leaves and stems, seeming to destroy all the vines. Yet the next spring, the vine buds burst forth with new life. Martin's donkey had inadvertently invented pruning.

In any case, in Austria, Saint Martin's Day traditionally marks the end of the current year's vintage and thus the beginning of a new vintage. During the fall of 2015, as I sampled the 2014 wines in the heuriger, I could tell that people were anxiously awaiting Saint Martin's Day so they could move on to the 2015 vintage, which by all accounts would be a great one.

Nowhere was that truer than in Carnuntum. Named for an ancient city on the Danube, Carnuntum was a Roman military fortress and one of the most important cities of antiquity, once boasting over 50,000 citizens, an enormous city population for that era. Significant ruins have been unearthed, including an amphitheater that once held 15,000 spectators, and a monument called the Heidentor, or Pagan's Gate, that are now part of a protected archaeological park. Of course, Carnuntum's importance is long gone. Its wine culture was almost completely destroyed during World War II. Now it is one of Austria's smallest wine regions, with no prestige DAC appellations and only about 40 producers. Yet many believe it's the very best place to grow zweigelt.

While its parent, blaufränkisch, is a show pony that wows the wine cognoscenti, zweigelt is much more of a humble workhorse. It was created in 1922, in a lab at the Klosterneuburg oenology school by Dr. Fritz Zweigelt, the botanist who crossed blaufränkisch and sankt laurent to create a grape that he originally called rotburger—not exactly the most appealing name to English speakers. After Dr. Zweigelt's death, the grape took its creator's name. Zweigelt for the most part is the sort of gulpable, everyday red that goes well with schnitzel and wurst and pretty much any meat tossed on a grill. The amount of zweigelt planted in Austria has increased more than 50 percent since 2000, to more than 16,000 acres, most of it sucked down in heurige without much contemplation.

But in Carnuntum, they make a special zweigelt-based wine called Rubin Carnuntum, which may be the grape's greatest expression. I visited with a young, slim and tanned winemaker named Phillipp Grassl, who drove me up to his prized Schüttenberg vineyard, overlooking a valley of slowly turning windmills stretching out toward Slovakia. "This area is like a windpipe," he said, as the air buffeted us. The wind apparently regulated temperatures and kept the grapevines dry during wet seasons. The question that producers in Carnuntum have been asking themselves is: Can zweigelt be a great wine? "We've been searching for 15 years for the ideal vineyard," Grassl said. "It's hard to have any kind of reputation because the appellation is so young. Some critics say it's too simple." My feeling is that, if you don't like a good classic zweigelt from Carnuntum, you'd better think about whether you like wine at all.

Grassl and I ate a lunch of Martinigansl, or Saint Martin's roasted goose, with knödel and red cabbage, at a wonderful local restaurant called Bittermanns, in Göttlesbrunn's tiny center. During the lead-up to November 11, Martinigansl is the specialty.

Pictures of cartoon geese appear all over rural towns during this season. Roasted goose, it turns out, pairs brilliantly with zweigelt—bright, juicy, and full of black pepper and blue fruit flavors, the perfect counterbalance to fatty meats. Nearly every great wine country offers a Tuesday night wine like zweigelt, a complement to the local Serious Wine. In that way, we can think of zweigelt similarly to dolcetto in Piedmont or perhaps petite sirah in California. "With zweigelt, a lot of winemakers go for the floral notes like Burgundy pinot noir," says Grassl. "But I want blueberry. I want dark fruit. I think there is a tension in this wine that's important."

Grassl told me that his father was a simple farmer who made wine along with raising livestock and growing other produce. "Every Thursday, they'd pack up the bottles and take them to Vienna to sell in the wine taverns," he said. "The wine was cheap and pretty bad. But it was little risk." After attending oenology school in the early 1990s, Grassl went to work in California, in Sonoma County, at La Crema, which was a subsidiary of giant Kendall-Jackson. The Sonoma experience was eye-opening for a young man from Göttlesbrunn. "There were all these professional technologies," he said, talking about the use of additives, stabilizers, clarifying agents, acidifiers and deacidifiers, and aggressive new-oak aging. "When I came home, I started out with a technical mind. I did everything I legally could to the wine. But eventually, I realized it doesn't matter what you do if the grapes aren't good." These days, like most good winemakers, he's more focused on the vineyards. Or, as a favorite contemporary winemaking cliché goes: *Great wine is made in the vineyard, not the cellar.*

After lunch, I walked around the corner to the cellar of Birgit Wiederstein, a small producer in her late 30s, who bottles her wines with whimsical labels and names like Zirkus Zirkus, Ein

Traum (A Dream), Frizzi Mizzi, and Die Diva (The Diva). One label was named after Rhea, daughter of earth mother Gaia. But, umm . . . Die Rhea might need a new label if it's imported into the United States.

Wiederstein, with her short bleached-blond hair, was less technical and more poetic in her approach to wine. She spoke about the grapes like people. "Zweigelt has a very nice, very simpatico way of being. He's flexible. He can chill or not. He's also someone that excuses mistakes. Blaufränkisch, on the other hand, is like an old man who lives on the mountain, who maybe doesn't like talking that much. And you have to sit with him a long time before he talks with you," she said.

But Wiederstein was committed to growing some extremely obscure varieties, including brauner veltliner, *brown* veltliner, which until recently was thought to be extinct. "It's not for money, I really just want to know what it tastes like," she said, noting that the vines wouldn't mature for a few years. She's also planted the Hungarian white variety furmint, in a nod to nearby Eastern Europe.

"Carnuntum was always the border to the Communist East," Wiederstein said. "This is all recent history. As a girl, I can remember, in the autumn of 1989, seeing people drive in from the east in their pastel Trabant cars. This was after Austria said they would not enforce the border anymore." She could also remember lying at night in the early 1990s, at her parents' summer house in the southern province of Styria, hearing the distant fighting in the former Yugoslavia. "And a long, long time ago, in the 16th century, the Turks came here and burnt down the villages. They did this like four or five times."

To this day, Carnuntum remains a geographical flashpoint as the trail of Syrian refugees trudges west, literally through the grapevines. "This past summer," Wiederstein said, "I was in the vineyard, and we saw groups of eight to ten Syrian people, just walking through the rows. We shouted to them, 'Can we help you? Can we get you some water?'"

As we gulped down Die Rhea, which name aside is an excellent classic zweigelt, we talked about how wine was so intertwined with history and the movement of human beings. Wiederstein began to tell me another story about that summer's growing season, and she was choosing her words very carefully. She asked if I remembered a horrific event from the previous summer, when Austrian authorities found a tractor trailer with 71 Syrian refugees inside, all of them—men, women, and children—dead from suffocation. "My car was only a few cars behind that truck when they found it," she said. "I went to get a palette of wine at the warehouse, and I'm wondering, why am I sitting in this traffic jam with lots of police around? And then I heard it on the radio." She added, "That's the news, not history."

Later that same evening, back in Vienna I was sipping another glass of zweigelt at a crowded wine pub. The barman and I were chatting about what makes a great zweigelt.

A drunk scruffy guy eavesdropping next to me guffawed, and said, "So you are American?"

I said yes and he asked, "Do you like our zweigelt?"

"Oh yes, very much," I said.

"So you like our zweigelt," he said, smiling with cigarette-stained teeth in a somewhat menacing fashion. "So do you know that zweigelt is a Nazi wine?"

"What do you mean?" I said, recoiling.

From the end of the bar, he grabbed a worn copy of that day's *Der Standard*, Vienna's largest newspaper, and then drunkenly tried to flip through the pages and translate for me. But I soon settled up and left.

Back at my Airbnb, I found the day's article online and let Google translate: "Some wines leave a bitter taste," Christina Fieber, the wine critic, wrote. She then railed against the fact that the zweigelt grape was still, in 2015, "inexplicably still named after its inventor." Besides inventing several grape hybrids, Dr. Zweigelt was, according to the article, a butterfly collector and "moderately talented amateur poet." More significantly, he was also a committed Nazi, joining the party as early as 1933 when it was still forbidden in Austria. After Hitler's annexation of Austria in 1938, Dr. Zweigelt was promptly promoted to head the Klosterneuburg oenology school, where he served until the end of World War II. "Besides the usual harassment of Jews he also delivered a student with connections to the resistance movement to the Gestapo," the article reported. Dr. Zweigelt died in 1964; in 1975, the name of the rotburger grape was officially changed to zweigelt, in his honor. It seems hard to believe now, but the name change happened during an era when Austrians were living by the societal myth that they'd been "the first victims" of Nazi aggression. It would be years before they collectively dealt with their legacy of Nazi complicity.

I'd always heard whispers that Dr. Zweigelt was "nationalist minded," but I'd never seen him confirmed as a Nazi until now. A queasy feeling developed in my stomach. So I'd come all this way just to learn that one of my favorite grapes was named after a Nazi. I was always recommending zweigelt to friends—the ideal wine for the backyard grill. Now, I could just picture a sensational, misguided headline in an American publication like, "Was zweigelt a Nazi grape?" I worried that this would have a similarly disastrous effect as the 1985 antifreeze scandal. I'd later learn that the Austrian Wine Marketing Board has had serious conversations about changing the grape's name. But what's their alternative? To start calling the grape by its orginal name, rotburger? I can't imagine a single American walking into a wine shop and saying, "Y'all got any of that rotburger?"

At that point I couldn't sleep, and so I opened a bottle of Grassl's 2013 Rubin Carnuntum, which was fantastic.

During the year following my visit to Austria, I discussed the issue of Dr. Zweigelt and his Nazi leanings via email with Grassl, one of the few winemakers willing to address it. In his thoughtful email responses, and then over wine when I visited him again the following fall, Grassl said, "I know the Wine Marketing Board would like to sweep this under the carpet. But they definitely made a mistake when they officially renamed rotburger in 1975. However, I think these days, maybe it's important to keep the name, to remember what happened in the past," he said. "Besides, if we change the name, then we'll also have to change the names of a thousand streets in Austria"—the ones presumably named after Nazi sympathizers.

The zweigelt issue popped up a number of times during my Austrian late-autumn sojourn. A few days later, when I visited Kamptal winemaker Willi Bründlmayer, in the town of Langenlois, we actually tasted zweigelt from Dr. Zweigelt's own vineyards. "I think it's crazy that 70 years later, someone puts these things out in the newspaper. Maybe it's a journalist who wants to make a name for herself," said Bründlmayer, with a sigh. "I don't know what will happen. Maybe we will have to change the name."

Dr. Zweigelt's zweigelt, however, was overshadowed by Bründlmayer's grüner veltliner and riesling, and especially his sankt laurent, perhaps the most finicky of Austrian red-grape varieties. Sankt laurent, named after Saint Laurentius, patron saint of cooks, is usually described as being like pinot noir, but more muscular. Which makes sense, since pinot noir is one of its parents. "Pinot noir had sex with some stranger and the child is sankt laurent," Bründlmayer said.

There is twice as much sankt laurent as there was a decade ago, but there are still only around 2,000 acres in Austria, so it remains obscure. Sankt laurent is notoriously difficult to grow. Winemakers often describe it as they would a crazy, troublesome lover. "I've always been a supporter of sankt laurent," Bründlmayer said. "But last year I thought, 'This might be the end of our relationship.'"

Part of the reason, like in many wine regions, is the changing climate. Bründlmayer says, for instance, that the average temperature in Kamptal has risen one full degree Celsius since the 1960s. "That's very significant, he said. "That means grapes ripen at least two weeks earlier."

For dinner, Bründlmayer pulled out from deep in his cellar a bottle of 1950 sankt laurent, covered in a thick layer of black mold that looked like fur. At dinner, we decanted the bottle and then tasted the 65-year-old, brick-colored wine. Incredibly, there was still freshness, with strange aromas of licorice and radish. This wine may have needed a cane or a walker to get around, but it was most certainly still alive.

I felt really lucky to be tasting this sankt laurent. For Austrian wine, 1950 is about as old as one might ever taste. Many older, prized wines dating to the Habsburgs were lost when the Klosterneuburg oenology school where Zweigelt taught was bombed in 1945. The decade following the war was obviously a chaotic time. From 1945 until 1955, Austria was occupied by the United States, the UK, France, and the Soviet Union. The wine regions of Kamptal and Kremstal were part of the zone governed by the Soviet Union. I'd been told by several winemakers that the Russians confiscated and drank a lot of the best wines during the 1940s and 1950s. "We don't have many older bottles here because the Soviet troops drank them," said Barbara Koller at Schloss Gobelsburg, just down the road from Weingut Bründlmayer. "I hope they enjoyed them." Martin Moser, winemaker at Hermann Moser in Kremstal, showed me a gigantic wooden door in his cellar, with an elaborate carved frieze, that was built in 1947. "The Russians wanted to drink everything," he said. "So my family tried to create this heavy door to keep them out."

During this era, around the town Langenlois, farmers didn't want the vineyards growing on the terraces high above the town. Bründlmayer's father bought up a lot of terrace vineyards for seven Austrian schillings per acre, equal to roughly 25 cents per acre in the 1950s. Many of those terraces are now considered among Austria's finest Erste Lage vineyards. That night, at dinner, we drank a wine from one of those vineyards, a 2004 riesling from hallowed Heiligenstein. It was agreed that this decade-old riesling was still just a baby—it had at least two or three more decades of aging potential.

We were joined at dinner by Willi Bründlmayer's son Vincent, some of his local buddies, and David Foss, a sommelier from New York and wine buyer at Anfora wine bar in the West Village. I've spoken several times in this book about "hip New York somms," and lest you think I've been using it as some sort of straw man, Foss fits the bill. And I use the term mostly respectfully. I have great admiration for people like Foss, who always has adventurous wines like Austrian neuburger or Georgian saperavi or biodynamic Chinon cabernet franc or Lebanese red blends on his list—yet engages his customers in plain language and with a great sense of humor.

Anfora is the Greek word for amphora, a reference to the clay vessels that have been used to ferment and store wine since antiquity. Foss is a fierce advocate of natural wines, as well as wines from Austria and Eastern Europe—so it makes sense he has bottles from winemakers like Roland Velich and Christian Tschida on his list. Because of Anfora's name, the bar is especially known for its selection of orange wine, which has become incredibly popular among the wine cognoscenti. "What exactly is an orange wine?" asked one of Vincent's friends at the table. Foss explained orange wine is white wine that's macerated and fermented with its skins. With this skin contact, it's essentially a white wine made like red wine, creating red-wine-like structure and an orangeish color—sometimes called amber, sometimes called *ramato* ("coppered") in the Italian region of Friuli-Venezia Giulia, along the Slovenian border.

Foss was in Austria brushing up on his Austrian wine knowledge for an upcoming appearance on Gordon Ramsay's television show *Hotel Hell.* He was asked to help improve the wine program of a crazy bed-and-breakfast in the Boston suburbs called the Vienna Inn. "This is supposed to be an Austrian restaurant, but they sent a photo of their cellar and it was all bottles of like Rioja, California zinfandel, and Chianti!" Foss and I hit it off and we hung out over the next several evenings in Austria, at one point drinking 1990s vintages of Mittelburgenland blaufränkisch at four A.M. at a wine bar called Pub Klemo in Vienna.

As our dinner with Bründlmayer rolled on, we opened a 2002 grüner veltliner from Lamm vineyard, yet another Erste Lage. This wasn't a baby, but it was still determined to be young. One of Vincent Bründlmayer's buddies, who was the son of Langenlois's mayor, held forth on collecting older cars. Apparently, he had quite a collection of old BMWs. As it turns out, my first car when I was a teenager was a beat-up 1980 BMW. My father was going through a phase of buying classic cars (he'd early purchased a prized 1964½ Mustang), and he and I split the cost, with money I made from my summer job. What drew 18-year-old Jason to a BMW is now a mystery, but I guess it had something to do with the Big '80s and the movie *Wall Street* and Gordon Gekko. The car ran so poorly that it was sold before I went off to college. In my self-delusional, revisionist history, I don't really see myself as the kind of guy that once drove a BMW. Anyway, at dinner with the Bründlmayers, I sheepishly mentioned that I once owned a 1980 BMW.

"A 318i?" asked the car collector. "Sunroof?"

"Yes," I said. "I think so."

"Ah," he said. "We call that one a baby. That's too young."

It was a curious exchange, measuring my taste in antique cars in the same way wine snobs might engage over old wines, with snap judgments and rankings and posturing. As I finished my 2002 grüner veltiner from the Lamm vineyard—still "young"—I thought about men and their hobbies, whether they be wine or music or art or sports or cars. I remembered the first time my father allowed me to drive his prized 1964½ Mustang convertible, which was all original and in mint condition. He made me understand that replacing even a hubcap would cost hundreds of dollars. As I tooled around with my friends that night, top down, speeding down a South Jersey country road, I hit a huge pothole. And then watched one of those very expensive hubcaps roll into the dark woods. When I drove back down the road the opposite way with my high beams on, I hit the same exact pothole and another hubcap flew off. As I ran into the woods after it, up to my shins in mud, having a panic attack as I felt around under bushes of poison ivy to retrieve the goddamned hubcaps, my friends all laughed hysterically. I never wanted the responsibility of driving that Mustang again.

Sometimes when I now drink an older wine, I think about my Uncle Bill—who died at age 80 as I was finishing this book. Bill was also an antique car collector, and he showed his cars at auto shows all over the east coast, winning numerous awards. But Bill wasn't drawn to flashy luxury cars; he collected Volkswagens. VW Beetles from the 1970s were mainly what he drove and showed, and he had a license plate that read "BUG." At Bill's funeral, there was a 40-year-old photo of his old yellow convertible bug at the beach, with about five of my cousins standing up in the back seat giving everyone peace signs. A younger Bill is sitting in the driver's seat looking cool, sunglasses on, his arm out the window, wearing a huge smile, not caring that sandy kids might mess up his upholstery. Bill knew how to be devoted to his hobby, but also not to take it all too seriously.

Volkswagen, of course, is an icon of 20th-century American pop culture, from Herbie of *The Love Bug* to Woodstock to the Beetle's huge relaunch in the 1990s. Few people seem to care anymore that the Beetle was developed by Hitler to be the "People's Car" in Nazi Germany, even though the idea of the "Volk" in "Volkswagen" was caught up in all kinds of racism and anti-Semitism. Like Volkswagen, I think it's probably too late to change the unfortunate name of the grape created in Dr. Zweigelt's lab. But I still hope that, one day, zweigelt or rotburger—or whatever you want to call the wine I love—has some modicum of Volkswagen's mainstream success.

The evening after our dinner with the Bründlmayers, both Foss and I were in Burgenland. We met Roland Velich for dinner at a restaurant called Taubenkobel, about ten minutes from Rust, where I once again ate Saint Martin's roasted goose. Velich brought with him a younger producer, Hannes Schuster of Rosi Schuster winery, whom he was collaborating with on a new project. Velich and Schuster are preserving an ancient winegrowing area in Zagersdorf, very close to the Hungarian border. Fossil grape seeds showed evidence that people may have been making wine in this place for more than 3,000 years.

Part of their collaboration involves producing a zweigelt, which they've chosen to label as rotburger. "Why?" I asked.

"Well, we're Hungarians. We used the name rotburger."

"Really?" I said.

"OK, why?" he finally said. "Why? Because Dr. Zweigelt was a Nazi. People say, 'Oh, he was a nice guy, let's not think about it.' But we all know he was a Nazi. 'We were the first victims,' people say. But 100,000 people turned out in Vienna to cheer for Hitler when he rode into town."

CHAPTER TEN

Gray Pinot, Blueberry Risotto, and Orange Wine

PURPLE FOOD, for most people, is weird. But one afternoon, at a restaurant called Pillhof in the Italian Alpine region of Alto Adige, I was served a curious violet-hued plate of a risotto, made with fresh blueberries. My encounter with purple *risotto ai mirtilli* was not exactly love at first bite, but with each forkful, as I wrapped my brain around the incongruous—yet, in the end, savory—flavors, my affection grew exponentially. Before I was halfway finished, my thoughts drifted away from the table. I was having a double Proustian memory of both the ripe, midsummer blueberries of my South Jersey childhood and my own clumsy, study-abroad attempts to master the art of risotto-making, namely to impress an Italian woman who remained woefully unimpressed.

Blueberry risotto might be a strange dish upon which to have a madeleine moment. But then Alto Adige is a peculiar place, a corner of Italy where a majority of the locals speak German, and they call the area Südtirol (or South Tyrol) and *not* Alto Adige. The town where I was eating the *risotto ai mirtilli*, Appiano, would be called Eppan by the German speakers of South Tyrol. In fact, many who identify as South Tyrolean still feel a pang of nostalgia for the Austrian Empire, to which they belonged until the empire collapsed after its defeat in World War I. An ambivalence, even hostility, to Italian ways lingers in Südtirol, dating to Mussolini's fascist program of Italianization, banning German language and culture, which one can imagine proved extremely unpopular here. Ironically, *Il Sole 24 Ore*, Italy's major business newspaper, regularly lists the Germanic towns of South Tyrol at the top of its annual *Borghi Felici* (Happy Village) rankings, trolling the hill towns of Tuscany and Umbria.

It's hard to imagine a more beautiful vineyard setting than Italy's northernmost winery, at the 12th-century abbey Abbazia di Novacella, or Kloster Neustift. There, at over 2,000 feet above sea level, they grow grapes like sylvaner, müller-thurgau, and kerner, a frost-resistant cross between riesling and a local grape called schiava that was created in 1929 and named after the 19th-century German poet Justinus Kerner, famed for such drinking songs as "*Wohlauf, noch getrunken*" ("Arise, still drunk"). I tasted wines in the old cellar where images of Austrian Emperor Franz Joseph were carved into the barrels. "You can feel the snow in the wines," said Dr. Urban von Klebelsberg, Abbazia di Novacella's winemaker.

On the day I was served *risotto ai mirtilli*, I was having lunch in Pillhof's courtyard with Letizia Pasini, the export manager for the winemaking cooperative in the village of Colterenzio, or Schreckbichl ("scary hill") if you prefer the German. "I hope you realize how different Alto Adige is from the rest of Italy," Pasini said, as we scooped and ate purple rice.

Pasini and I had spent the morning wandering steep vineyards and then tasting, in Colterenzio's glassed-in tasting room, never losing sight of the snowy mountains. The brilliant sunshine in Alto Adige is otherworldly, with 300 sun-soaked days each year that, combined with the altitude, create a microclimate perfect for what may be Italy's greatest white wines. As a cooperative, Colterenzio makes wine from the grapes of 300 small growers working less than 750 acres. Unlike in the rest of Italy, where cooperatives are synonymous with cheap and mediocre wine, the village cooperatives in Alto Adige are highly regarded. Pasini suggested that cooperatives worked better here because of the "German mentality" of the region. "People are a little more focused here. There is more strictness here. And more professionalism," she said. "But we still have a little bit of the Mediterranean flair."

I'd heard a similar sentiment about South Tyrol's "German mentality" many times during my visits to Alto Adige, often less diplomatically. "Things just work better up here than they do in Rome or Milan," I'd been told years before by an export manager at Alois Lageder, one of the best-known estates in Alto Adige. "That's not complaining or bragging. That's just a fact."

Colterenzio, like most of the wineries in Alto Adige, grows many different grapes, from well-known international ones like pinot noir, pinot blanc, and pinot gris to obscure ones like lagrein and schiava. And then there's gewürztraminer, which falls somewhere in between. As we tasted, Pasini proselytized the wonders of pinot blanc, which is called pinot bianco or weissburgunder here, depending on your language. "My mission has been to have pinot bianco become the next pinot grigio in the United States," she said. "Many people are getting sick of hearing about pinot grigio. I mean, come on. It's been going on for how long?"

All the pinots—pinot noir (black), blanc (white), and gris (gray)—are mutations from a single variety. Pinot grigio is simply Italian for pinot gris, or *gray* pinot, because it happens to have mutated with grayish-pink berries. (Even the German speakers in Südtirol don't call their money grape *grauburgunder*.)

Pinot grigio's path to popularity is mind-boggling. Between 2000 and 2010, the amount planted in Italy increased by more than 60 percent, to 42,000 acres, roughly mirroring prosecco's explosive growth. Often neutral, bland, and lemonade-like, pinot grigio is the wine of choice for summer beach drinking and cougars on the prowl. The occasion for me feels like straight from the fridge, maybe after mowing the lawn, cold like a Bud Light. How to explain the continued popularity of insipid pinot grigio in the States? Does it balance America's dreams of la dolce vita with its commitment to safe, middle-of-the-road choices? Perhaps it represents the Subway Tuscan Chicken Sub of wine. Anyway, it is consistently the biggest rip-off in the wine shop. I regularly see people who complain about spending over $10 for a bottle walk into a store and drop $23.99 on a bottle of Santa Margherita. In fact, Santa Margherita pinot grigio, for years, was the top-selling wine over $20 in the United States, with more than $25 million in annual sales at its height.

Santa Margherita, as it happens, comes from Alto Adige—always in Italian; the Austrian heritage of Südtirol doesn't really fit the Santa Margherita marketing image. It's hard to imagine now, but at one time pinot grigio was a little-known variety. According to legend, Santa Margherita became Americans' favorite Italian white because of one night in 1979, when importer Anthony Terlato had an epiphany over a pinot grigio. "I felt that the well-known Italian white wines of the period, such as Orvieto, Frascati, Soave, and Verdicchio, would never sell for more than $5 a bottle," Terlato wrote in his 2008 memoir, *Taste: A Life in Wine*. "I returned to find a white varietal [sic] that could command a higher price in the high-end restaurants."

On his first night in Milan, Terlato was served a pinot grigio at dinner and his life changed. The next day he drove to a town called Portogruaro in Friuli, close to where the pinot grigio was from. At dinner that evening in his hotel, he ordered every pinot grigio on the restaurant's list, 18 of them. After tasting all 18 along with the owner, they agreed that the Santa Margherita was the best. The very next day, Terlato drove north to Alto Adige, to the palazzo of Count Marzotto, Santa Margherita's owner, and made the count an offer. Within days, the count signed a ten-year agreement with Terlato to export the wines. "I was about to introduce a practically unknown varietal to the American market, considerably above that of any other popular

Italian white wine," he wrote in his memoir. I don't know if it was dumb luck or Terlato had a knack for knowing exactly what the American palate wanted, but the rest is history. Santa Margherita split from Terlato in 2015, but during their 36 years in business, that wine made Terlato a boatload of money. Now entire wine shop shelves are devoted to a previously obscure Italian white variety.

Terlato wrote that he'd eaten a "dish of pasta in a simple tomato sauce with some fresh basil on top" on the fateful day that he tasted 18 pinot grigios in Portogruaro. I wondered what would have happened if he'd ordered the blueberry risotto that Pasini and I were eating that afternoon in Pillhof's courtyard. If he had, I wonder if he'd have decided that, say, pinot bianco or gewürztraminer—both of which we were drinking—would have been the white wine that would have brought him fame and fortune. But then gray pinot would have suffered the cruel fate of remaining unknown and unloved by Americans.

Pinot blanc, and not pinot grigio, is the everyday wine of Alto Adige—if you ask for a white wine there in a bar, you'll most likely be served a pinot blanc. The grape was brought to the region in the 1850s by the Austrians. "Alto Adige was the main producer for the monarchs in Vienna," said Judith Unterholzner, who works at Cantina Terlano (or Terlan in German), another village cooperative. Many believe that pinot blanc has great aging potential, and it's hard to disagree when I taste a bottle like Terlan's Vorberg Reserve, of which I tasted the ripe and creamy 2011 and the deep, nutty 2002. Both could have easily been mistaken for aged chardonnay from Burgundy.

Then there is gewürztraminer. One of wine's immutable truths is: *Gewürztraminer . . . People love it or hate it!* It's hard to think of another wine that's both reviled and cherished in equal amounts. When Lettie Teague, the *Wall Street Journal*'s wine columnist, wrote about a gewürztraminer tasting she'd conducted with friends, she called it "one of the most polarizing tastings I've staged in a while."

Growing up in Jersey, everyone had that one buddy who was, you know, a little too much. You know the type: He's loud, wears a little too much cologne, shows a little too much chest hair, wears a flashy watch and maybe a gold chain, tips people from a wad of dollar bills. When you're out with this guy, he can be cringe-inducing, and he's difficult to mix with certain friends, some of whom despise him. However—and it never ceases to amaze me—he still manages to charm over a surprising number of people with his overbearing act. Plenty of people simply love the guy. I often think of gewürztraminer as sort of like this buddy.

If you don't want to think of gewürztraminer as a swarthy Jersey guy, consider this description of gewürztraminer by a Viennese winemaker I met: "It's like when you have a classroom full of quiet, well-behaved children, but you have this one rowdy child that's very loud. That loud one is gewürztraminer!" Surely, you get the picture.

In any case, I happen to love gewürztraminer. Yes, it can be both overpowering and lacking in subtlety. Yet there is nothing like its flamboyant aromatics of rose and lychee and clove, and when the alcohol and sweetness are held in balance with even a dab of fresh acidity and chalky minerality it can be an absolutely thrilling wine. In seemingly impossible pairing situations, gewürztraminer is your buddy. Spicy Thai food, check. Stinky cheese, check. Smoked salmon, check. Peanut sauces, check. Blueberry risotto, *certo*. I always recommend it as the go-to bottle for that most American of autumn feasts, the 3,000-calorie Thanksgiving dinner. With its mash-up of contrasting flavors—turkey, yams, green bean casserole, brussels sprouts, cranberry sauce, pumpkin pie—there is no more daunting task of wine-and-food pairing. But gewürztraminer carries it off.

Still, many wine lovers remain unconvinced. I'm not sure which is more of a turnoff for them, the bold flavors, the fact that gewürztraminer is almost always 14 percent alcohol or higher, or simply the name—once again, fear of the umlaut. "When Americans cannot pronounce a name, they tend to not stay with a wine. Even if they like the wine," Pasini told me as we sipped Colterenzio's gewürztraminer. Perhaps if we call the grape by its Italian name, traminer aromatico, it might get more traction?

About 20 minutes from Pillhof is the idyllic Alpine village of Tramin (or Termeno in Italian). Tramin gives its name to a grape with many mutations. Like heida in Switzerland, the name traminer is synonym for savagnin, the godforsaken grape I've talked about in an earlier chapter. Gewürztraminer literally means "spicy traminer" and is believed to be a mutation. Traminer (or savagnin) is so ancient that it's likely the parent of pinot noir, sauvignon blanc, grüner veltliner, and others.

But whether the village of Tramin is the origin of the grape is unclear. The place name Traminer was definitely used to denote good-quality wines since at least the 13th century, but those wines were probably blends of local grapes. The origin of the actual traminer grape has been disputed by German wine historian Christine Krämer, who asserts that it came from southwest Germany. Whoever first cultivated the grape and named it traminer was likely trying to capitalize on the fame of wines from Tramin.

"The mayor of Tramin is always telling people that this is the home of gewürztraminer. But I'm skeptical. I think it comes from Germany," said Martin Foradori Hofstätter, winemaker at J. Hofstätter in Tramin. "But that's OK if it comes from Germany. Because at least it doesn't come from France." In any case, Hofstätter said, "Gewürztraminer grows perfectly right in front of my cellar door, so I focus on it."

Since it's grown in more than a dozen countries, gewürztraminer can be considered an international grape. For instance, Alsace, the region of France that borders Germany, is perhaps the most famous place where it's grown. Yet with a little less than 30,000 acres worldwide, gewürztraminer is still not on most wine drinkers' radar. One reason might be because much of the world's gewürztraminer is produced sweet, with significant residual sugar. "When the average American wine lover hears the word *gewürztraminer*, 99.9 percent of them say, 'I don't like sweet wine,'" Hofstätter told me. Yet what makes Alto Adige's gewürztraminer unique is that, for the most part, it's completely balanced and dry, with more subtle floral and fruity notes.

In Tramin, I also visited with Karolina Walch, at the highly regarded winery Elena Walch, founded by her mother. "We really believe in gewürztraminer as our variety," Walch told me. "And for us, it is always dry." We tasted her intense gewürztraminer, as well as her excellent Castel Ringberg single-vineyard pinot grigio. When I expressed my surprise at how serious and thoughtful the pinot grigio was, she said: "Yes, well, for us as a region, certain pinot grigios have ruined the reputation of pinot grigio." Walch told me that the gewürztraminer was actually the winery's bestseller within Italy. "Italians love aromatic whites," she said.

How does one convince Americans to also love aromatic gewürztraminer? For some time, I've believed that what gewürztraminer really needs is a rebranding. Why not? People and companies reinvent themselves all the time. Even food. Maybe gewürztraminer could be like kale, the lowly vegetable rebranded as a superfood. Or maybe like sriracha, which for decades was just a humble Thai sauce, until American fast-food places turned it into a mayonnaise. Or maybe like pumpkin spice or salted caramel or matcha tea or any number of flavors that no one thought about for years, and are now ubiquitous.

As it turns out, someone has beaten me to the rebranding idea. A few years ago, Cantina Tramin, the village's 300-grower cooperative, launched a campaign called Gewürztraminer Renaissance, which aims to show just how versatile and food-friendly the wines can be. While in Tramin, I visited Cantina Tramin and talked rebranding as I tasted with the winemaker, Willi Stürz. "Many people still think gewürztraminer is heavy and sweet and complicated. But it's actually very easy drinking," said Stürz.

With Gewürztraminer Renaissance, Cantina Tramin collaborated with chefs and sommeliers from nine different restaurants in Italy, Germany, Denmark, Japan, Thailand, and the United States. The goal is to dispel the notion that gewürztraminer is difficult to pair. Dishes ranged widely, from saffron risotto to fettuccine with smoked trout to pork shoulder and chanterelles to sashimi to scallops with spring onions. The campaign boasted a web page and a 35-page book of recipes.

"For us, gewürztraminer is our most important grape. But we feel a prejudice. Too many times a person says, 'Oh gewürztraminer, I don't like it,'" said Cantina Tramin's communication manager Günther Facchinelli—who has perhaps the most perfect Alto Adige / Südtirol name ever.

"It's very difficult to make good gewürztraminer," Facchinelli said. "It needs hot days. It needs sun. It needs freshness. If you do a tasting of gewürztraminer, there's a big range of good and bad. It can be an intense aromatic experience. An explosion."

At one point, later in our tasting, I tried to explain my personification of gewürztraminer as being like my old buddy from Jersey. Stürz and Facchinelli looked at me skeptically and didn't seem to understand. "We prefer to call gewürztraminer a woman. A diva," Facchinelli said.

Perhaps this idea explains the cheesy poem on the product page of Cantina Tramin's most prestigious gewürztraminer, Nussbaumer:

A sarong of gold silk on bare skin,
grilled shrimp on the beach at sunset.
Honey glances breathing sigh.
Rose on roses, drink with my eyes the song of fire.

"Don't you think she's a woman?" Facchinelli asked.

Perhaps that's the magical rebranding gewürztraminer truly needs: To transform my old hairy-chested, gold-chained buddy into a beautiful sarong-wearing diva.

In my mind, I began to conflate my discussions about gewürztraminer and the strangely enjoyable blueberry risotto. However much as I like to experiment, I'm also neurotic about tradition and authenticity. In years of traveling around Italy, I'd never seen a risotto made from berries, or any fruit. But several people in Alto Adige assured me that the dish was typical in the Alps. I eventually got a chance to speak with Daniel Sanir, Pillhof's chef. "I like to make something new," Sanir said, with a shrug. I asked if it was difficult to convince Italian-speaking diners to give blueberry risotto a try. Sanir told me it could be a challenge, but that the dish usually wins people over. "Sometimes the customers think, 'I don't know, this is strange with the berries.' But then they eat it and 90 percent of them like it a lot."

I was intrigued enough that I looked up several recipes for strawberry, apple, and blueberry risotto in the classic, bestselling cookbook *Il Cucchiaio d'Argento* (*The Silver Spoon*), essentially Italy's *The Joy of Cooking*. I searched online and found a spate of fruit risottos in Italian cooking magazines from the late 2000s. Yet as I delved deeper, asking around to Italian friends, I soon learned that the topic of fruit risotto was a fraught one. Most shut down the discussion immediately: "Fruit risotto? Preposterous!" some said, grimacing and holding their stomachs. When I pressed further, several acknowledged that berry risottos had a minor moment of popularity a few years ago, even being shown on TV cooking shows. "But I've never tasted it," those people said, defensively. "It sounds very strange." My friend Stefano grew so incensed when I raised the topic of fruit risotto that he shouted *"Mi fa cagare!"* ("That makes me poop!"), a common vulgar slang expression meaning "That's awful!"

Amid such mixed messages, I arrived at the conclusion that fruit risotto is definitely a thing that exists; but it's the sort of thing that most Italians won't talk about. Perhaps it's sort of like how David Hasselhoff's musical career is for Americans. So maybe it makes sense that they speak German in South Tyrol. Though even there, I faced evasiveness. I initially emailed IDM Südtirol, an agency that promotes the region's products and tourism, to ask about fruit risotto. The communications person replied, somewhat defensively: "I'd suggest 'risotto with speck' which is what reflects our local cuisine." I had better luck with a food writer named Yrma Ylenia Pace, an editor for the lifestyle magazine *La casa in ordine*, who published a recipe for risotto with wild berries on her blog *A Fiamma Dolce*. This recipe actually won an award at Milan's Expo 2015, sponsored by a large rice company, for "most beautiful risotto with Italian ingredients." Ylenia Pace, who lives in Alto Adige's neighboring Alpine region of Trentino, confirmed that berry risotto was a real thing in the Dolomites.

Most of the blueberry risotto recipes I found called for local, Alpine red wines, which I also explored while in Alto Adige. While pinot grigio may be a household name, and gewürztraminer divisive, Alpine reds like lagrein and schiava may currently represent the height of obscurity. "It's a great challenge for Alto Adige to get a reputation in the world for our reds," Karolina Walch told me.

"These are wines for specialists," said Hofstätter, who recounted his first business trip to the United States in the 1990s. "It was easier to sell ice to the Eskimos than to sell a bottle of lagrein." Still, for many years, the reds like lagrein from South Tyrol were some of the most prized in the German-speaking world. Charles IV, the 14th-century Holy Roman Emperor, declared lagrein to be among the greatest wines.

Of all the Tyrolean grapes, red or white, lagrein seems to me the one with the most untapped potential, with only 1,200 acres under vine. With its lovely inky color, great acidity, and soft tannins, lagrein is much lighter than the jammy, overripe, super-concentrated, or oaky reds we see too much. There's a freshness, a savory and mineral core, and often a bit of pleasant funkiness. That attractive edge of the rustic is what I love most. "I call this . . . *wine*. We talk too much about wine. We should enjoy more wines like this, a wine with edges," Hofstätter said. "But our goal is too keep too much of the rustic out of the wine." In good lagrein, you can feel the tension between rusticity and sophistication.

I visited Castel Turmhof, which the extended Ticfenbrunner family tree has owned since 1675. But wine was made here long before that: An estate called Linticlarus existed in Roman times, and records show that the Cathedral of Trento was ordering wine from here as early as the 14th century. In 1857, eight different wines from Turmhof were sent to the first Imperial Austrian agricultural exhibition in Vienna, where they won high praise from the Imperial tasting committee. Bottled under the Tiefenbrunner label, its most famous wine is a müller-thurgau grown at over 3,000-foot altitude, called Feldmarschall von

Fenner. Yet while the reputation of Tiefenbrunner's whites is well established, its reds were the revelation for me. I particularly was drawn to the lagrein reserve, which seemed to be inching into Serious Wine territory with its ageworthiness.

Christof Tiefenbrunner, of the latest generation to run the winery, also poured me his schiava, a unique, light-bodied, bright, tannic red that's almost rosé-like in color. Schiava is called vernatsch in local German slang and was once the everyday grape of Alto Adige. Even today there's more than twice as much schiava planted as lagrein. Tiefenbrunner said that as recently as 40 years ago, it made up 70 percent of the winery's production—but now is less than a fifth. "The problem for the grape is that the same vineyards that are good for schiava are also good for pinot noir. So the producers are ripping out the schiava," Tiefenbrunner said. Pinot nero, as they call it here, is taking over Alto Adige in the same way it does nearly everywhere.

Tiefenbrunner isn't the only one who extolled the everyday virtues of schiava/vernatsch. At Cantin Terlano, Judith Unterholzner told me: "We have a golden rule here in Südtirol. From the morning til noon, the old people are drinking pinot bianco. But when the church bell strikes noon, they immediately switch to schiava." Schiava, I can say with firsthand experience, is a dangerous wine if you are trying to stop at one glass.

You're beginning to see schiava on a few of the more innovative wine lists in the United States, often under the name trollinger, which is what the grape is called in Germany. I believe this sort of bright, almost-pink, but structured red—like maybe grignolino from Piedmont or listán negro from the Canary Islands—is the kind of wine that could explode in the United States. To me, it's the next logical step for an adventurous drinker who has fallen in love with rosé, or even someone who enjoys lightweight pinot noir. But sommeliers and retailers will tell you that in the United States there is a resistance to light reds. As Teague, in her *Wall Street Journal* column, once wrote: "To many wine drinkers, a light red is anathema, a wine defined by an absence"; one of Teague's friends told her, straight up, "A light red sounds downright un-American." For those of us who love schiava (or vernatsch or trollinger) this Trump-like response elicits a heavy sigh.

Near the end of my stay in Alto Adige, I visited Castel Sallegg, in Caldaro (or Kaltern, if you prefer the German). Since 1851, Castel Sallegg has passed from Archduke Rainer Joseph of Austria, Viceroy of the Kingdom of Lombardy-Venetia, to the Princess of Campofranco, to today's owner, the Count von Kuenburg.

Castel Sallegg sits in the warm microclimate surrounding Lake Caldaro and reds grow especially well here. In a sunny courtyard, amid orange trees and birds chirping, after a tour of the count's historic cellars, I tasted through the Castel Sallegg portfolio with Matthias Hauser, the winemaker. "This scene is almost too beautiful," I said. "I wonder how it will affect our tasting?"

Hauser laughed and joked, "Actually, yes, these wines are crap. But you won't really know since it's so nice and lovely."

The wines were actually excellent, a fact that had nothing to do with the scenery. The bold licorice-and-blackberry-kissed lagrein had that amazing telltale tension. The Bischofstein schiava was the ideal kind of "un-American" light red for daytime drinking—crisp, edgy, full of savory and dark fruit flavors, austere but surprisingly pleasant. It was the kind of wine that makes you throw away the cork, grab a plate of speck or prosciutto, and forget about work for the afternoon.

If I were a latter-day Anthony Terlato, drinking this schiava in the courtyard of Castel Sallegg, I might have had a similar epiphany. American wine drinkers' palates are swiftly changing. They want bolder, more complex whites and reds that aren't so big, jammy, and oaky. They want everyday wines like that schiava, with *drinkability* as the key word. Even if many people would find this wine "un-American" at the moment, a guy like Terlato might be able to look slightly into the future. If I were that kind of guy, I'd surely make Count von Kuenburg an offer he couldn't refuse.

After my stay in Alto Adige, I drove three and a half hours east through the Dolomites, past Lago di Santa Croce where I'd first tasted that timorasso at Dolada restaurant, then down into the prosecco zone, past Vittorio Veneto, where Umberto Cosmo and Cinzia Canzian push the limits of glera, then past Portogruaro, where Terlato did his legendary tasting of 18 pinot grigios. Finally, I entered the region of Friuli-Venezia Giulia, and drove until I reached Gorizia, up against the border with Slovenia. I was so close that when I took a taxi to a casino suggested by the hotel clerk, I was actually driven a few minutes across the porous border into Slovenia. Just like Alto Adige, Gorizia was controlled by the Habsburg Empire until World War I, when it was annexed by Italy.

I especially love the hard-to-pronounce Friulian reds made from indigenous grapes like refosco dal peduncolo rosso and schioppettino, the latter of which nearly disappeared in the 20th century but has been resurrected in a familiar Italian narrative. Pinot grigio also grows in abundance in Friuli. Yet on this visit, I was specifically focused on the native Friulian white varieties, particularly ribolla gialla and friulano. Even more specifically, I was interested in those white grapes being turned into orange wine—or ramato, "coppered," as they often call it in Friuli. Orange wine can be made from any white grape, but anyone who's been an adventurous wine drinker over the past decade has surely observed the proliferation of orange wines from here, where Italy and Slovenia meet.

About 15 minutes north of Gorizia, in the town of Oslavia (or Oslavje in Slovene), I visited two neighboring winemakers, Stanko Radikon and Josko Gravner. Both Radikon and Gravner began farming organically and using noninterventionist natural winemaking methods in the 1980s, and are early pioneers of the burgeoning natural-wine movement that's become such a force in the wine world today.

Gravner's epiphany came during a ten-day trip to California in 1987, tasting more than a thousand wines. He didn't like what he tasted. "When I returned home, I said, 'I don't like the way wine is going. Wine isn't wine anymore.' I had to take a different path from wines that are made in laboratories. When you learn there are more than 300 additives in wines, you don't want to drink those wines anymore. Wine loses its music, its poetry, when it's made in a laboratory."

The first thing Gravner did was tear out all of the chardonnay and pinot grigio he'd planted. In their place, he planted ribolla gialla vines. "Ribolla gialla has been around here for 1,000 years, and it's only grown here." Then he started using native, wild yeast, and stopped using sulfur or filtering his wines. "Once a wine is filtered, it's just a beverage." In the late 1990s, he made a pilgrimage to the Republic of Georgia, where he learned how to make wine in qvevri, and then brought the clay vessels back to Oslavia. In his winery, there were no steel tanks or barrels, only a solitary, dimly lit open room with the tops of the qvevri that are buried underground. Gravner basically dumps his harvest into the qvevri and lets the grapes express themselves into wine.

"These wines are difficult to understand," Gravner told me as we tasted. "They disturb people." It was indeed hard to pin down what was going on with Gravner's ribolla gialla: nutty one moment, waxy the next, suddenly salty, moving to oxidative, then mushroomy, on to warm tangerine and dried apricot, then finally back to nutty. My mind kept changing, minute by minute as we tasted—I loved it, I hated it. One moment I thought Gravner was a genius, the next . . . maybe the emperor was wearing no clothes. The man had even invented his own wine glass, almost like a short, round rocks glass, that he believes best accentuates his wines. I made the mistake of calling the wines "orange" and he scolded me. "We don't call them orange," he said. "We call them amber."

Down the road, Stanko Radikon was less dogmatic than Gravner, though he followed similarly strict natural-wine practices. He told me that his father had encouraged him to plant the old, indigenous varieties. "You have to plant ribolla gialla, because that's the local variety and it will make you happy." But Radikon also worked with pinot grigio, chardonnay, and sauvignon blanc. Unlike Gravner, Radikon didn't use the Georgian qvevri, but rather huge old Slavonian oak barrels for his orange wines, and he let the wines age for as along as five or seven years.

Perhaps his finest wines, though, came from the grape friulano, which is called ravan across the border in Slovenia, and sauvignonasse in southwest France, where it originated. In fact, how the name friulano came to be used is typically topsy-turvy. In the 19th century, friulano had been called Tokaji in Italy and then, from the 1930s, tocai friulano. At the turn of the 21st century, however, Hungary complained to the European Union about the Italians' use of tocai friulano—in an effort to protect Tokaji, its prestigious appellation. In 2008, Italian winemakers were forced to officially change the grape's name to friulano. Ironically, during the same era, the Hungarians also complained to the EU about how winemakers in Alsace, France, called gray pinot by the name "Tokay." Since 2007, the name Tokay is banned and the Alsatians must bottle the gray pinot as *pinot gris*.

"Orange wine? It's just a red wine made from white grapes," he said, with a shrug. Radikon was similarly cheeky about his labels. He called one of his wines Orange, in English. His most famous orange wine, made from 100 percent friulano, he labeled as Jakot—Tokaj backwards, a poke at the EU and Hungary. Jakot was also a confounding wine: It was fresh and full of vibrant acidity, but one moment a pine forest, the next moment a menthol cigarette, the next a piece of dark chocolate, the next a handful of almonds. Did I like it? Does it matter? When I told Radikon that I saw Jakot on a restaurant wine list in Philadelphia for $98, he winced. "Too much," he said.

Radikon died of cancer in September 2016 as I was writing this book. Eric Asimov, in a *New York Times* obituary, wrote that "his wines epitomized a level of beauty and truth sought by wine lovers every time they open a bottle." I only spent a couple of hours with the man that afternoon. But I clearly remember what he told me as we finished tasting and he bid me arrivederci: "These are the wines of the future, because they're natural."

"But thou hast kept the good wine until now."

— JOHN 2:10

III.

SELLING OBSCURITY

CHAPTER ELEVEN

Waiting for Bastardo

SEVERAL YEARS AGO, I found myself in Logroño, Spain, in the heart of Rioja wine country, at an event called the Digital Wine Communications Conference. This was soon after I had finally given up my cocktail column and left the *Washington Post* to pursue wine (after two years of insufferably lobbying my editor had been to no avail). The Digital Wine Communications Conference was previously called the European Wine Blogger Conference, but the event had apparently changed its focus to accommodate the burgeoning segment of "content creators" who were neither wine journalists nor wine industry marketers, but rather existed in a gray space of consultancy and freelance assignments. The best parts of the conference were the tastings of aged Rioja and other Spanish wines. I mostly avoided the seminars, such as "Challenges of a Global Brand in Communicating Socially." To be honest, I was a little desperate to find freelance work, and I had made a connection with an editor of a British drinks magazine who was looking for a columnist. Mostly, he and I just hung around Logroño's amazing tapas bars eating embuchados, patatas bravas, and Galician-style octopus to soak up all the wine.

On the final day of the conference, Wine Mosaic hosted a grand tasting panel, moderated by José Vouillamoz and his coauthor Julia Harding, titled "Native Iberian Varieties." *Wine Grapes* had just been published, and this was the first I'd heard about Wine Mosaic and its work rescuing native grapes. I arrived a little late, and was seated in the back, next to an older British wine writer named Robert Joseph, whose panel "What Is Wine Communication and Who Are the Wine Communicators?" I'd sat through. Joseph fancies himself a snarky truth-teller, and his PowerPoint presentation mostly scolded us "snooty" writers for not talking more about popular grapes like pinot grigio and mass-market brands like Yellow Tail, Barefoot, Skinny Girl, Cupcake, and Apothic. He excoriated us for not being populists and told us we should be more like the film critic Roger Ebert because he praised superhero blockbusters like *Spiderman* and broad comedies like *Meet the Fockers*. Finally, he said our work didn't really matter, since "Bill from the office," among his social media followers, will probably be a more influential wine communicator than any of us. It all seemed rather cynical: The idea that most consumers were so hopeless about wine that you should just pander to them.

At the grand tasting, Vouillamoz and Harding presented ten wines. I was particularly captivated by the obscure and ancient Portuguese varieties: vital, alvarelhão, touriga fêmea, jampal, antão vaz. I'd already been to Portugal a half-dozen times, but I'd never heard of any of these grapes. The fact that antão vaz, fresh, crisp, and redolent of ripe apricots and honey, and jampal, smoky and full-bodied and delightful, were both Portuguese *white wines* made them even rarer.

Just as I was beginning to slip into a meditative tasting frame of mind, Robert Joseph began speaking loudly—to me—in the same snarky tone from earlier. "How much are we spending to support this charity?" he said. At first I tried to ignore him. People sitting around us could hear Joseph's heckling, and I didn't want them to think we were together. But he kept on, louder. "We've now spent 40 mintues on the same wine!"

"Don't you like the jampal?" I whispered, looking around uncomfortably, hoping to tamp down the heckling. It did not work.

"Hmmm. Maybe we need to spend 40 more minutes to experience more fully its not-very-evident flavor."

I didn't know what to do. Joseph is an eminence of sorts in the British wine writing circles. He was the wine correspondent for the *Sunday Telegraph* for 16 years and has written 28 books. His conference bio said he was both a Chevalier du Tastevin, the secret society for Burgundy drinkers, as well as a member of the exclusive Bordeaux brotherhood, the Commanderie du Bontemps du Médoc et Graves. Those seemed like pretty highfalutin memberships for a guy who hectored other wine writers to be populists and talk about Yellow Tail and Barefoot. But what did I know. Besides, who the hell was I to shut him up? I had no standing in the world of wine. Maybe eminent wine writers were supposed to be cheeky assholes?

"I really like this one," I said, motioning to the delicate, young, low-alcohol alvarelhão. "Maybe it's a little subtle. But if you take another sip and give it a think . . ."

"Ha!" he cut me off. "If you were stuck in a prison cell, you could probably taste water and think about it for a very long time too!"

I was done with the Digital Wine Communications Conference. I left the room before they poured the touriga fêmea. And it wouldn't be until a couple of years later, in Switzerland over raclette, that I would finally meet Vouillamoz.

I was still thinking about this episode the next day, when I arrived at the airport to find out my flight home from Bilbao was delayed seven hours by a terrible wind storm that shut down several European airports. I spent five and a half of those seven hours stuck in a line of hundreds, while two overwhelmed workers at the Lufthansa desk ever-so-slowly attempted to reroute 300-plus passengers. As the line trudged forward, I watched the departures board helplessly as flights left, one by one, for Paris, for London, for Madrid, for Lisbon, all connections that would have gotten me home. I had an important meeting in the morning, and then my son's first soccer game, which I'd committed to coach. As the hours passed, I knew I would miss both. By the time I reached the front of the line, there was no way across the Atlantic until the next day, and I was assigned an evening flight to Frankfurt. I was given a handwritten voucher for a hotel, and another voucher for a free dinner.

Travel has the singular ability to turn the most banal events into heightened drama. This can be great for writing. This can also be not-so-great for situations that call for less drama. In the grand scheme of my trips and travails, this was all relatively small potatoes. During many years of travel, I've been a passenger on two planes that nearly crashed, a bystander caught in the midst of political demonstrations that turned into violent riots, and a victim of a few felony crimes. This travel disruption wasn't even in the same league as the erupting Icelandic volcano that stranded me in Italy. My typical response to a seven-hour flight delay would usually just include some cursing and a few useless, angry phone calls to the airline, eventually giving way to heavy sighs and then drinking.

When I arrived at Frankfurt airport, it was dark and rainy, and a taxi took me to a hotel in the middle of an industrial park in a suburb called Mörfelden. After checking in and explaining to my son that I would not be home in time for soccer, and hearing my boss's dismay at my absence, I slumped down to the hotel's overlit restaurant and grabbed a menu. I was a wreck. This trip was supposed to help kickstart a career in wine. But it hadn't. In fact, it made me think all this wine stuff wasn't for me. In any case, I badly needed some comfort food, and the first item that called out to me was Wiener schnitzel. Why? I don't know. Maybe I was channeling my mother's old veal parm back in Jersey. Maybe it just felt like the opposite of the tapas I'd been gorging on for days. In any case, I ordered the Wiener schnitzel and presented my voucher. The stern waiter sneered and pointed over to a pathetic buffet in the corner: some stale rolls, a congealed soup, and a platter of rubbery chicken that had been sitting out for hours. This, apparently, was the Lufthansa Stranded Passenger Special that my voucher covered.

I waved the waiter back over. "Please sir," I said. "Please. I've had a very long day, and what I really need is to eat this Wiener schnitzel."

"It's 21 euros," he said. "That food over there is free."

I had kept mostly cool and zen all day long, but I suddenly had the totally irrational urge to scream or cry. "Look, I don't care what I have to pay for it," I said, my voice rising. "I just need you to bring me this Wiener schnitzel. Right now. Please." Something in the stern waiter's demeanor seemed to change, empathy washed over his face. He nodded, wrote my order, and whisked away the menu. A few minutes later, he brought a plate with the schnitzel. And along with it, a bottle of Rheinhessen riesling.

"Sir," he said. "I am so sorry. I cannot honor the voucher for your meal. But please. I asked my manager, and he said I could pour you this riesling in exchange for the voucher." I thanked him quietly, and averted my eyes, blushing.

I ravenously tucked into that schnitzel, and took a long drink from the wine glass. It wasn't the greatest schnitzel or riesling, but for some odd reason, my eyes started to well up and tears ran down my cheeks. These were tears of frustration, but also very much of embarrassment. I'd suddenly realized that the heightened travel drama happening in my own head had selfishly put this poor waiter in a tricky professional spot. He'd only wanted to make sure I clearly understood that I was passing up free food. Surely I wasn't the first agitated, stranded passenger he'd faced at this airport hotel, and surely there'd been misunderstandings and complaints in the past. He didn't need any trouble from his manager on this lousy night in Mörfelden.

Meanwhile, I'd turned the moment into some kind of angsty epiphany. Perhaps only a ridiculous person who reads too much travel writing would think of Ryszard Kapuściński while eating schnitzel alone in an airport hotel, but I was reminded of that line in Kapuściński's masterpiece, *The Soccer War*, when he writes: "There is so much crap in this world and then, suddenly, there is honesty and humanity."

This whole episode, the sudden-but-quiet shift, from banal to dramatic, from stupid to beautiful, from melancholic to comic, from insignificant to profound was something I found happening often during my wine journeys. Wine revealed itself in mysterious ways. The German cultural critic Walter Benjamin once said, "I would like to write something that comes straight out of things like wine from grapes." I believe Benjamin wrote that while he was high on hashish, so perhaps I misunderstand him, but I wanted to do the same. In any case, when I was writing about cocktails, I was sort of a loudmouth—"acidly opinionated" had been some praise—like Robert Joseph. It was a persona that worked for that cocktail realm, with its flashy bartending personalities, flamboyant distillates, and wild, not-always-factual storytelling about liquors. Certain details didn't matter so much. And certainty about cocktails was easier to come by.

But wine was different, I was learning. It was impossible to know everything about wine. Plus, there was always someone, some expert, telling you that you're doing it wrong. So wine always introduced a note of doubt. You could ignore it, you could roll with it, but the doubt was always there. That was what drove people insane about wine. At the same time, I was starting to realize that everyone, even the most important gatekeepers, who presented themselves as so super-confident in their wine judgments and opinions were probably themselves seized by doubts. Even Robert Parker once said: "My personal philosophy is, you can be sure of nothing." In that light, I could empathize with an older wine writer like Robert Joseph. For someone who's staked their whole career on serious opinions on Serious Wines, I could understand how destabilizing a tasting of obscure varieties might be, especially if they'd never tasted them before.

As I ate my schnitzel and drank my riesling, I thought back to the Portuguese wines that Joseph ridiculed. If this guy was truly a populist wine crusader, a consumer advocate guiding people to good-value wines, then Portuguese wines should really be on his radar.

I'm always speaking with people who are fixated on a quixotic quest to find that "great bottle under $10." I often get frustrated with this mythical idea of the under-$10 bottle, because it's almost impossible to find one that offers quality and drinkability and isn't full of additives or chemicals. I'm almost always advocating that people bump up at least a few bucks into the $15 to $20 range. A $9.99 wine—poorly made and artificially enhanced—can offer bad value just as often as some overly ambitious $29.99 wine can. Portugal, however, is one of the exceptions, a country that regularly produces wines under $15 that offer honest-to-goodness value. Which is why their lack of presence in the United States continues to surprise me.

I eagerly look forward to the meteoric rise of Portugal's great-value wines, either from the Douro Valley or from lesser-known regions such as the Alentejo, Dão, Bairrada, or Setúbal. My wait has been very much in the vein of *Waiting for Godot*, and it has been going on for over two decades now. But I remain patiently optimistic. What I appreciate best about Portuguese winemakers is how they've stuck mainly with their ancient grapes—there are nearly 80 varieties that are indigenous to Portugal.

I have some hope that the younger generation will finally take up Portuguese wines. Nuno Vale, the director of Wines of Portugal, told me that millennials consume 43 percent of the Portuguese wines sold in the United States (well above the market average of 26 percent for all wines). "Right now, drinkers are looking to explore new varieties," Vale said. "They're curious and want to have new experiences." Of course, at the same time, Vale hopes that Portugal can shift its perception from "cheap" (fizzy $9.99 vinho verde) to "value."

So why haven't more consumers embraced Portuguese wine? Perhaps they're simply scared away from unfamiliar grapes, such as arinto, encruzado, touriga nacional, or trincadeira (aka tinta amarela). An Iberian grape that a consumer might know, tempranillo, is called by completely different names in Portugal: tinta roriz in the north or aragonez in the south. Similarly albariño is called alvarinho here, and mencía, more famous in Spain's Bierzo and Galicia, is called jaen in Portuguese.

Even a hipster favorite like trousseau is grown here under the name bastardo—in fact there is nearly ten times as much trousseau/bastardo grown in Portugal than in Jura, the mountainous part of France that's made the grape fashionable. In Portugal, bastardo is mostly used in port blends, but it's beginning to be bottled on its own. Of course, a grape named trousseau is likely much more appealing than one called bastardo. Improbably, the Portuguese grape I see the most on wine lists right now is baga, with its mossy, forest-floor-and-wild-berries funkiness. The authors of *Wine Grapes* actually call baga "controversial" and say it's "either loved or loathed" and makes both "the best of wines and the worst of wines." Only an adventurous wine drinker is going to fall in love with a grape like that. Regardless, I still have faith. I'll wait for consumers to embrace bastardo and baga.

Strange grapes aside, though, there's also a lack of geographical awareness. If you say the word "Portugal" to some people, their faces go blank. "Portugal? So flying down ol' South America way, huh?"

My family, however, has a strange, deep connection to Portuguese wines. For us, Portuguese wine will always conjure memories of the Bedford Rascal, a ridiculously tiny sea-foam green minivan that my parents, in a clear display of insanity, rented in the summer of 1992 for a long family driving tour through Portugal. This was right after my college graduation, and how my parents—who'd never visited Europe before—settled on this itinerary still remains a mystery.

They thought they'd scored a deal from a fly-by-night rental car company. A whole van for the price of a subcompact! The rental-car equivalent of a $9.99 wine! It didn't take long for the excitement to wear off. In the airport parking lot, my father scratched his head and mumbled, "This looks like a beer can on wheels." We learned some interesting things about automobile design and nomenclature from the Bedford Rascal. We learned, for instance, that it is possible to invent a vehicle that cannot comfortably seat five small-to-average-size people and still, with a straight face, use the word "van" to refer to this vehicle. We learned that it's possible to invent a vehicle that has no hood, and to stash the battery, oil gauge, and other important fluids underneath the back bench seat, and still not be compelled to call this vehicle a go-cart. Finally, it is possible to equip a highway-legal vehicle with a choke lever, similar to what you might find on your lawnmower at home.

For a glorious week, we puttered around the countryside at top speeds of 40 miles per hour. After fidgeting with the choke lever, my father realized that if he wedged the lever out in a certain fashion, we could increase our top speed to a full 45 miles per hour. The downside of this method was that the next time we tried to start the minivan, the engine would not turn over. Every morning, my brother and I would have to push my father down some type of incline to jump-start the car. Luckily, we stayed at a number of hilltop pousadas. Every day of our vacation began with my mother and youngest brother running down a steep cobblestone lane, luggage in hand, and leaping into the back of the moving Bedford Rascal.

As we left Lisbon, we slowed to pick up a hitchhiker, a cute old weathered man with a cane. But as we pulled onto the shoulder, and the man saw what he might have to ride in, he vigorously waved his arms and shooed us on our way with his cane.

He must have known something we didn't, because about a half hour later, as we crossed the bridge over the Tejo River, the Bedford Rascal sputtered. We had to stop in a town called Setúbal, an industrial fishing town on the coast of a peninsula south of Lisbon, because my father decided the Bedford Rascal "just needed a rest." We decided to eat dinner, randomly chose a traditional, tiled restaurant near the port, and entered. In 1992, we may have been the first American family to have ever set foot in this restaurant, and when we did, the only sounds were from a soccer game on the television.

At first, the server who took our order was gruff, especially as we pantomimed and I tried out laughably bad Portuguese phrases, which I'd been learning from cassette tapes. The unsmiling owner brought over a carafe of astringent and nearly undrinkable red wine that smelled like a wet dog. With great contempt, he asked, "You are British?"

"No, no," I said. "We're American."

Suddenly, something happened that I have never experienced in Europe since. Upon learning we were American, the owner's face brightened into a big smile, and he called his wife and son over. "They're Americans," he said, seemingly relieved we were not part of the hordes of boozy, sunburned Brits that take over Portugal in the summer.

The whole mood quickly changed. Complimentary plates of meats and cheeses came out. The owner whisked away the carafe of cheap plonk, and his son brought over a bottle from the cellar and uncorked it for my father to taste.

"Wow. Tell him this is wonderful," said my dad, with surprise. I believe when the bill came it was about $4 a bottle.

In 1992, the only Portuguese wine most Americans knew was Mateus, the sweet fizzy rosé known for its goofy flask-shaped bottle. Mateus (which, incidentally, was Saddam Hussein's favorite wine) accounted for almost half of the table wine exported from Portugal in the late 1980s. But what we were drinking was a far cry from Mateus.

What the owner had opened and poured for us was a bottle of Periquita, Setúbal's most famous wine, produced by José Maria da Fonseca from mainly castelão, a native variety that is Portugal's most widely planted grape. Twenty-five years later, Periquita (imported by giant Palm Bay International) is still one of the most widely available Portuguese bottles that you can find in the States—and one that will still only set you back around $10. On that day with my family, it was indeed wonderful: dry, tannic, smoky, dark, rustic. It was the unmistakable taste of old Europe in the glass. We drank two bottles and forgot all about the troubles of the Bedford Rascal, and then stayed the night at a nearby pousada.

The next day we drove through the Alentejo. Meaning literally "beyond the Tagus (River)," the Alentejo begins a little over an hour east of Lisbon, and for years was overlooked as tourists passed through on their way to the Algarve beaches. By contrast, the Alentejo, the poorest, most sparsely populated part of Portugal, is a beautiful, melancholy place with blistering, sun-baked summers and chilly winters, vineyards and olive groves, castle ruins and hill towns. Since driving through in the Bedford Rascal more than 25 years ago, I've returned to the Alentejo numerous times. It hasn't changed very much.

These days, it's standard among wine people to say that the Alentejo is Portugal's most experimental wine region, more in line with New World wine regions like Australia or Argentina. Much of this is based on the Alentejo's most famous and large producer, Esporão, with its forward-thinking Australian winemaker David Baverstock. Esporão's modern-design labels were among those that appealed to the young millennials in the tasting I did while preparing to teach The Geography of Wine.

But if you travel the Alentejo, a different picture emerges. The back roads of the region are lined endlessly with trees stripped bare of bark. Undressed from the branches down, dark trunks cozy up tight to the road, as if gathered together in an attempt to escape the vast openness of the sunny green and gold plains. These trees are cork oaks, and their bark has been peeled away for making wine corks. Portugal is the world's leading supplier of cork, and the Alentejo is cork's epicenter. Cork takes a great deal of patience and faith to produce. A tree planted in 2018 will not be able to produce usable cork until about the year 2058. Yet thousands of new cork oaks were planted in the Alentejo during the 1990s, even as the use of non-cork closures such as screw caps and other synthetic wine closures gained momentum. Think for a moment about a stubborn country that plans four decades ahead to grow a product that may or may not be replaced in the near future by a screw cap. When you've thought for a moment about that, you will be in the right frame of mind to experience the Alentejo. And Portuguese wine in general.

—

Back in Mörfelden on that lousy night, I had a private laugh at how distraught I'd gotten over the damn schnitzel. Now I felt foolish, though I was certainly enjoying the comfort food. I am aware that some of you reading this may be offended by the mention of this dish, made of veal. Around my more enlightened friends it makes me vaguely uneasy to admit that I eat veal. I usually downplay it significantly. "Yeah, I'll eat veal maybe once in a blue moon," I might say. Of course, people have legitimate reasons for not eating veal. Commercial veal producers over the years have not done veal lovers any favors by the way they raise and treat their animals, and so I do look for good, conscientious producers as much as possible. And I very much respect why some people do not eat veal. But the truth is: I love veal. Breaded veal cutlets are among my fondest foods of memory, and one of the only dishes I've consistently eaten from childhood up through today. I also love the Italian version of schnitzel, *cotoletta alla Milanese*, which I first ate as a 19-year-old studying abroad in Lombardia. In fact, just like the origin of grapes, there is some debate over whether Wiener schnitzel or cotoletta alla Milanese came first. Milan, like a lot of Europe, had fallen under the rule of the Austrian Empire. For years, legend had it that famed Field Marshal Joseph Radetzky had brought the recipe back from Italy after his battle triumphs in the mid–19th century.

Maybe schnitzel makes me think a little bit about tradition—the good and the bad—about how empires fall, and about how, amid change and upheaval, small anachronistic joys persist. Veal is not comfort food for everyone, but many of us look upon it fondly, just as many people feel a nostalgia for the strange and incoherent Habsburg Empire. Veal, like wine, seems like a silly thing to argue over. Yet on the other hand, it seems like exactly the kind of thing we should always be arguing over.

The waiter reappeared and said, "Is everything OK, sir?"

"Yes, yes," I assured him. "Thank you very much. Everything is quite OK."

CHAPTER TWELVE

The Same Port Dick Cheney Likes

THAT RAINY NIGHT IN MÖRFELDEN wasn't the first time I'd I deconstructed my life in order to reinvent it. I did it first about a decade before, after I'd quit a gig as a dining critic for a glossy city magazine in Philadelphia. When I say I quit my restaurant critic gig, that is likely an understatement. My quitting was in the style of those Buddhist monks in the '60s who lit themselves on fire in protest—except my protest wasn't over anything important like an unjust war or civil rights. It was just the latest in a long line of incredibly self-sabotaging moves. In any case, I was at loose ends, but I'd landed a travel assignment to write about Porto, and focus on visiting the city's famed port lodges. I asked my brother Tyler to come with me to Porto, many years after our family trip in the Bedford Rascal, to drink port for a few days. This traveling on assignment to drink things, and then writing about the traveling and the drinking, would very soon become my calling in life. But I didn't realize it quite yet.

Tyler and I arrived in Porto on a quiet Sunday evening, after a rain, and the slick, cobblestone alleys along the Douro River glistened in the lights. Across the Douro, in the suburb of Vila Nova de Gaia, we could see the huge illuminated signs of the famed port wine companies and their Anglo names—Cockburn, Burmester, Taylor's, Dow's, Graham's, and the shadowy, black-caped logo of Sandeman.

The next day we visited Graham's port lodge, on a hill high above the Douro River. We tasted the highly regarded 2000 vintage, which we loved. On the tour, we checked out a special cellar devoted to famous guests of honor who've visited throughout the years. Each special guest was invited to throw the contents of a glass of 20-year-old tawny at one of the "appropriate" wooden vats—individually labeled "The Minister," "The Ambassador," "The Emperor," "The Sportsman," and so on. Some of the luminaries so honored include cricket star Graham Gooch, big-game hunter Simon Fletcher, the Duke of Argyll, and Major General Sir John Noble Kennedy, colonial governor, in 1954, of British Southern Rhodesia, known now as Zimbabwe.

It was not surprising that the atmosphere inside many of the port lodges seemed an anachronistic slice of the British Empire. The British have dominated the port trade since the late 17th century, when importing wine into England from enemy France was banned. The British had to drink something, and so fortified wines from Portugal were sought after. By 1703, the Methuen Treaty between Portugal and Britain reduced the duty paid, and port wine's popularity soared. In 1756, the Douro Valley wine region's boundaries were demarcated, establishing it as the third-oldest official wine region in the world. The wine would be shipped down the Douro River to Porto where it would age. In 1806, the Portuguese government deeded land so that the British port producers could build the Factory House, a gentleman's club that still exists to this day, along with other edifices like the Oporto Cricket and Lawn Tennis Club and the Oporto British School (the British called Porto "Oporto" for some reason). So long story short: Just like Bordeaux or Burgundy or Champagne, Port has serious old-school credentials as a Serious Wine.

When I think about port, I think of my earliest, clumsy attempts at seeming—with requisite air quotes—"sophisticated," or at least "fancy." Back in my early 30s, port seemed like the fast track to connoisseurship. "I'll take a glass of the '66 Fonseca," I'd say to a waiter as everyone else was simply ordering dessert.

I'll admit I've always been kind of insufferable. But I did grow fond of port, and it did end up being the first wine I truly came to know.

Later, sitting in the clubby tasting parlor of Taylor, Fladgate & Yeatman port lodge, Tyler and I worked our way through glasses of white port, several late-bottle vintages, a colheita, a 20-year-old tawny, and the 1994 vintage, considered by many to be the greatest port of the 20th century. While we tasted, a peacock with feathers unfurled roamed the lovely English gardens outside. Just to be contrarians, we decided that we liked the 20-year-old tawny better than the greatest vintage of the 20th century—though this was a pose, mainly because neither of us wanted to believe ourselves to be the kind of man who desired a $300 port. "This tawny is more of an everyday port," I said, ignoring the fact that it was still over $50 a bottle.

Sipping on the tawny, I wandered into the library to see the modest display of famous people who drink Taylor's port. What I learned was rather shocking. Though we likely agree on nothing else, apparently Dick Cheney and I share the same taste in port. Next to a photo of Cheney was a yellowing article from the October 6, 1990, edition of London's *Daily Express* bearing the headline "Cheney's port in a storm." The reporter commented that, while our troops in Saudi Arabia during the first Gulf War had been deprived of booze, "I can disclose that America's Secretary of Defense has not been so abstemious in his alcohol consumption." Cheney had humped his own bottles of Taylor's port into the desert. Lest we draw too many conclusions from all this, let me point out another documented devotee of Taylor's port: Fidel Castro. Some day, maybe we can achieve such common political ground outside the civilized parlor of a port lodge? One can only hope.

I could go on with my tasting notes on the Taylor's tawny or the Graham's vintage, tell you about mellow ripe fruit, aromas of raisins and meatiness, hints of plum and blackberry. But now that I've inserted elements of class and colonialism and geopolitics into my port tasting, do you care? Can I ask you to put that out of your mind? If we can, what does that say about our taste? About our critical faculties? What does it say if we can't? It's the same question I asked that night in Vienna when I found out about Dr. Zweigelt.

When I returned to the tasting room, my brother was just finishing the last of his tawny.

"By the way," I said. "Hate to break it to you, but we like the same port that Dick Cheney likes."

"Damn," he said. "But this shit is so good."

After our visit to Taylor's we wandered back down toward the river to eat lunch. We wanted to eat something authentic. But the most authentic thing to eat would be tripe, which Tyler wasn't having any of. The people of Porto are known throughout Portugal as *tripeiros*—"tripe eaters"—which explains the inordinate number of market vendors who specialize in stomach lining. According to legend, in the 14th century, Porto's residents selflessly gave away all their high-quality meat to the Portuguese explorers who discovered the New World, leaving everyone at home with only tripe to eat. Much as I've ruminated, I'm entirely uncertain what lessons to draw from that parable.

We saw the stone walls and dozens of cured and smoked hams hanging above the bar at a place called Adega e Presuntaria Transmontana and decided a big lunch was a good idea. We ordered cheeses like Serra da Estrella and Terrincho and a sausage made from a rare black boar, served on a wooden plank. We drank an amazing, bold, full-bodied red from the Douro Valley, the kind of wine I keep waiting to take off in the United States. It was made from the same historic grape blend as port—touriga nacional, touriga franca, tinta roriz, tinta barroca, and tinta cão.

I had to admit that, despite my assignment to drink port wines, I preferred the dry, non-fortified Douro table wines like that one. That made me uneasy, conflicted. I figured I was supposed to love port as a Serious Wine. Yet even on that day, I could tell that my love for port was waning. Like most everyone else's, it seems. Sure, you can find port in most wine shops, at least for now. But sales are in serious decline. By some industry estimates, port sales have dropped by more than 25 percent over the past decade. If port keeps trending this way, becoming more and more underappreciated, it may actually freefall into obscurity. It's like watching another wine empire collapse. For the future children of millennials, it could end up being just as foreign and strange as wines with umlauts.

Still, even though I rarely drink port, I give the port producers a lot of credit. For centuries, they've preserved their native grapes. Surely, without port, varieties like touriga nacional, tinta barroca, tinta cão, and others would have been ripped out long ago. Now that they've been protected, we all can eventually enjoy the rise of the Douro Valley's great-value red table wines that I keep waiting for.

The waiter encouraged us to order the house specialty: posta mirandesa, which we were told was a veal steak cut from the hindquarters of young calves that ate only natural food and roamed free in the mountains of the remote region of Trás-os-Montes. Yes, yet another veal dish. It occurred to me that, just like port, people consumed a lot more veal in decades past. Fifty years ago, Americans ate, on average, something like four pounds of veal per year. Now? The average is less than a half pound.

As we awaited our veal steaks, Tyler and I recalled another saga from the family drive through Portugal in the Bedford Rascal. This happened at the end of our trip, once we'd made it to the Algarve. One night, Tyler and I grabbed the keys and, following the usual jump-start, headed straight to a nearby pub.

We left the Bedford Rascal idling outside and soon found ourselves drinking with a Portuguese man, who told us we should continue on to the next village where there was a disco that was popular with young people.

I knew that this man was not using "disco" in the John-Travolta-finger-pointing-shiny-ball sort of context. But Tyler did not understand this, and he refused to go to any disco. (To this day I'm not sure why—perhaps he feared being forced to wear a white suit and do the Hustle.) A showdown ensued. Outside, Wilson family tensions reached an all-time high, escalating into the first fistfight between my brother and me since grammar school. All this over who would be captaining the Bedford Rascal and whether it would go to a disco. As we began to grapple, the minivan's engine coughed and sputtered to a halt. It took every able-bodied man in the pub to push us to a start.

Our last stop was a few days later in the coastal town of Sagres. Unfortunately, the hotel in Sagres was not on a hill. The next morning, we had a flight to Madeira from the town of Faro, and my father insisted on us awakening before five A.M., to allow at least four hours for the drive. My father hadn't slept all night, fearing the worst, which happened. The Bedford Rascal would absolutely not start.

We pushed and pushed, but finally ran out of incline. We left the car in a ditch near our hotel, and called a taxi. Our driver showed up in a big yellow Mercedes. We, and our luggage, all piled in comfortably, and he drove us at high speeds to the airport in Faro. The trip took about an hour. At the airport, we slammed the keys on the rental car counter in disgust. The bored-looking rental car people didn't seem the least bit surprised and dragged on cigarettes with dangling ashes. We drew them a map of where the Bedford Rascal now rested.

Years later, at the Adega e Presuntaria Transmontana in Vila Nova de Gaia, Tyler I both laughed about our infamous altercation as the waiter brought our huge slabs of posta mirandesa. I had never seen or tasted veal like this—garlicky, vinegary, a hint of spicy pepper, thick but beautifully tender meat. This was nothing like the insipid, pale, unethical veal you find at home. Veal may not be an everyday pleasure anymore, but it's amazing when it's prepared like this.

After finishing lunch, we both ordered glasses of 30-year-old tawny, the color of an old brick. Just like the monster steaks, the tawny was utter decadence: thick, viscous, generous, with layer upon layer of rich walnut, chocolate, and caramel. Perhaps port wine, sweet and heavy and old-fashioned, is sort of like the veal of the wine world. After sharing the veal and the family saga of the Bedford Rascal with my brother, that three-decades-old port still felt very important, still very much a Serious Wine.

Several years after that my trip with my brother, in the spring of 2009, Porto's producers unanimously declared 2007 to be a vintage for port. Vintages are one of the things that keeps port relevant as a Serious Wine, and it only happens a few times each decade, and so the declaration is a very big deal, at least among the shrinking number of people who get excited about port. The previous vintage had been declared two years earlier, in 2003. Since 2007, there have only been two vintages, in 2011 and 2014. In April 2009, a handful of these people were invited to the Four Seasons restaurant in midtown Manhattan for a preview tasting of these 2007 vintage ports from 11 highly regarded producers.

I was invited along too. Perhaps it was because, by that time, I was working as a cocktail columnist. Unlike most wine writers, I was probably one of the few reporting on port, which was just being touted as a cocktail ingredient at trendy mixology bars. Or perhaps my invitation had been a mistake. In those days, I wasn't rushing out to fork over $200 for a bottle of anything, unless maybe that bottle contained a genie who would grant three wishes. In any case, I was surely not going to pass up this event, even if it was just to make my brother jealous. I was so overexcited, in fact, that I ended up getting my dates wrong and took a bus up from Philadelphia on the Friday beforehand. When I presented myself at the Four Seasons, the maître d' gave me a withering look. Someone behind him shouted that the port tasting was on "*Lunedì!*" So I returned on Monday.

"A very special occasion on the wine calendar," according to the official black book of tasting notes each person received when we arrived. It was a Who's Who of port wine brands, including Graham's, Taylor's, Croft, Fonseca, and Quinta do Noval. Producers set up shop around the perimeter of a private dining room. Each company offered a pour of their 2007 vintage, as well as at least one older vintage port—classics such as the Graham's '70, the Taylor's '77, and the Fonseca '85—presumably a best guess at what the 2007 will mature to be as it ages. Islands of spit buckets stood in the center of the room. Waiters circulated with trays of hors d'oeuvres.

Our host was Anthony Dias Blue, the bon vivant who used to be *Bon Appetit*'s wine and spirits editor and now runs high-end tastings like these. Resplendent in a double-breasted blazer, Blue pointed out that—unlike his home in southern California, where people come to tastings "in shorts and flip-flops"—the crowd at the Four Seasons looked very dapper and formal. I think that meant most of us (including me) had bothered to throw on a jacket and some of us (not me) had worn a tie. Roughly three-quarters of the attendees were older than me by at least two decades.

After an hour or so of tasting, Blue introduced Rupert Symington, whose family has made port since the 19th century and owns Graham's, Dow, Quinta do Vesuvio, and Warre's ports, and he approached a podium in the corner of the room. Symington is the sort of youthful, aristocratic-seeming British chap that Americans are instinctively drawn to.

Symington spoke for all the producers. "Declaring a vintage port is a difficult decision," he said. "Only about three times a decade do we decide to stake our reputation on a single year." Perhaps, as Symington himself suggested, it seemed a bit discordant to be introducing a traditional luxury product like vintage port at the Four Seasons on a springtime Monday afternoon, in the midst of a global economic collapse? "Well," he said, "we've made the wine already. To hell with the crisis!"

"The Ivy League of vintage port" is what Symington called the 11 port houses on hand. "These all represent the safe bet," he said. "These wines will last the test of time, which is really the point." Which meant Symington's intended audience was the type of person who planned to buy up cases of this stuff, and then cellar it for a few decades.

Several of Symington's cousins joined him at the podium, and then the session was opened for a Q&A. There was a polite question from a *Food & Wine* writer about the 2007 growing conditions and what made them extraordinary (it had been an unseasonably cool July and August, followed by intense heat in early September). Then, after no one else ventured a question, a very tall man with a British accent shouted, "When will you be announcing the 2007 prices?"

"I don't know," Symington said. "Let's just say there's a debate. At the moment, the jury is still out." What was left unsaid: How does one price a vintage port in an economic climate where port sales are plummeting?

Even the price of top-end vintage ports had stagnated. Sure, a renowned vintage (like the Taylor's '77 or the Graham's '70) will fetch $200 to $300. But how did that compare to, say, a highly rated Bordeaux from 2009 or 2010, which could go for more than $1,000? Top-shelf cognacs sell for $5,000 or more. Next to those, vintage port begins to look almost anachronistic, another era's "luxury" item—sort of like a big Cadillac, or a pool table in your rec room, or a fountain pen. As critic James Suckling was moved to ask, in the January/February 2009 issue of *Wine Spectator*: "Who drinks vintage port anymore?"

Port's reputation rests on the vintages, and there is a mystical quality to the years widely considered to be the best: 1947, 1955, 1963, 1966, 1970, 1977, 1985, 1994. This emphasis on vintages, of course, offers poseurs an easy path to connoisseurship. Humorist Andy Borowitz explored this phenomenon in a 2000 essay in *Food & Wine* on vintage port. "No other beverage in the world has the power to make me look like such a complete and utter know-it-all," Borowitz writes, adding that he learned everything he knows about port in six minutes, simply by memorizing three dates and four producers. "Unquestionably, when people hear me toss off 'Fonseca '70' or 'Taylor's '63,' they assume I've come to port only after honing my mastery of wine, cigars, antiques, and Greek mythology."

Vintage port still only makes up about 2 percent of the world's port supply. All port wine grapes are grown in one of the world's oldest appellations, the Douro Valley, and port is stored and aged in Vila Nova de Gaia. Explore the other 98 percent of port and you'll find all sorts of interesting wines: tawny ports, aged from ten to 40 years in oak barrels; colheitas, which are barrel-aged ports taken from a single vintage; late-bottle vintages that have been left in the barrel for four to six extra years. Most of these are ready to drink once you buy them, and you can drink them over several weeks, unlike with vintage ports, which mostly age in the bottle, and need to be consumed within a day or two of opening.

In a *Decanter* magazine article on declining port sales right before the vintage declaration, Francisco de Sousa Ferreira, director of Portuguese wine giant Sogrape, admitted: "We need to reinvent ourselves." In the United States, that meant attempting to convince American drinkers to use port as a cocktail ingredient rather than drinking it after dinner. This was an era when you saw a lot of newspaper and magazine articles saying, "You really don't need to be an elderly Brit to enjoy port" and making the case that port should be hipper.

This was the same playbook used by sherry, the other fortified wine from the Iberian peninsula that was traditionally enjoyed by elderly Brits, and was lagging in sales. For sherry, the cocktail path had been successful, and hipsters everywhere seemed to embrace the drier fino, manzanilla, and amontillado expressions. Most port, however, was sweet wine and had less application in the types of dry, new-wave cocktails that had become popular. During this era, I wrote an article suggesting port as an ingredient in such cocktails as the Philadelphia Scotchman (with apple brandy, orange juice, and ginger beer), the Princeton (gin, port, orange bitters), or the Perfect Pear (cognac, port, muddled pear). I even suggested mixing a 40-year-old tawny, white rhum agricole, and a dash of bitters, a drink that I called—wait for it—the MidLife Crisis, "garnished with a red convertible." (I'll be here all week, try the veal.)

Unfortunately, the attempts at popularizing port also meant the development of a number of lower-quality, lower-priced bottlings designed in a quixotic attempt to recruit newer, younger, or "nontraditional" drinkers (translation: Americans, possibly women, who drink pinot grigio, prosecco, and just tried an Aperol Spritz). Croft had recently released Croft Pink (at $17, it's "the first pink port ever!") and encouraged drinkers to enjoy it chilled or on ice. Warre's had also rolled out its "affordable" Otima ten-year Tawny (around $26) with a marketing pitch encouraging consumers to chill it and serve it as an aperitif. "Can't finish a bottle in one night? Otima is perfect for the everyday or occasional drinker, as it preserves its quality for up to three months after opening when stored in the refrigerator." Noval Black, a $20 ruby, aged about three years, from the 400-year-old winery Quinta do Noval, was the very best of these "cocktail" ports. In a telling move, Noval hired famed mixologist Jim Meehan, rather than a sommelier, as its brand ambassador.

In my role as a cocktail writer, I'd actually visited Rupert Symington at the Graham's Port Lodge in Vila Nova de Gaia ("the day after the Queen's birthday," I was told). After we toured the family's cellars, checking out dusty, rare bottles from the early 20th century, Symington halfheartedly touted the company's Six Grapes label, which was being marketed to the cocktail crowd for about $15. Symington spoke about how "the States is one of the few places you could do grassroots marketing." Port knowledge in the United States, however, remains painfully low. Even a lot of high-end restaurants with expensive vintage ports on their dessert menus don't understand that they're meant to be consumed within an evening or two of opening the bottle. "I'd walk into a restaurant," Symington said, "and they'd have ten bottles of vintage port, all open, none of it decanted, sitting under the lights."

It was all a bit depressing to listen to a truly distinguished connoisseur like Rupert Symington describe running around like a huckster from restaurant to bar in American cities "with a big bag of booze," demonstrating pairings of cheese and chocolate with port—all part of that grassroots marketing effort. "Actually," he said, with a heavy sigh, "I do feel port is best enjoyed by itself."

A grand lunch was served in a lush dining room that overlooked the Douro River and the city of Porto. I was invited to join a party thrown in honor of the sales staff of the Portuguese duty-free airport shops, who'd been selling a lot of port to tourists. There was no $15 Six Grapes on the menu. Symington broke out a Warre's 1937 Colheita. It was, as a British gentleman might say, extraordinary. To think that the wine I was sipping had gone into a barrel the same year as the Hindenburg burst into flames, the year Guernica was bombed in the Spanish Civil War, the year the world's last Bali tiger was shot by a hunter in Indonesia. It was like a time machine, and I will never forget it. The Warre's 1937 Colheita, unlike the cocktail ports, was served at room temperature with no ice cubes.

In the private dining room at the Four Seasons, after the presentation and the Q&A, everyone went back to tasting in earnest. I sampled all 11 of the 2007s on offer. Perhaps here I should mention that 2007 was indeed an excellent and shocking year, producing port wines with a unique freshness and acidity as compared to other recent vintages, but with tannins that provide great structure but are still silky smooth. Several—such as the complex, floral-eucalyptus Graham's, the fresh-and-clean Quinta do Noval, the jammy-and-anise Smith Woodhouse, and the rare Taylor's Vargellas Vinha Velha (only 200 cases would be available)—were easy drinking right then, so young, which is generally unheard of.

As I tasted, I kept flashing back to something Symington had told me on my visit during the unseasonably cool summer of 2007 that helped create this vintage: "We're trying to convince people to serve vintage ports younger." Even the *Wine Spectator* critic James Suckling wrote, "I think you can start enjoying them after about six to eight years." Maybe this vintage could be the first to test that idea?

As the event wound down, I bumped into the tall man with the British accent who'd shouted the question on price at Symington. He introduced himself as Julian Wiseman, an investment banker and port enthusiast around my age who lived in New York and hosted a website called *The Port Forum*. "So what did you think?" I asked him.

"Why are we even drinking these today?" Wiseman said. "They're not going to be ready for 20 or 30 years?"

When I disagreed and suggested I could see drinking them a little earlier than two decades from now, Wiseman—even though he was a likeable guy, a good bloke—immediately sensed I was not a kindred spirit. "If you open even the best of these in six years," he protested, "it will never taste the way, say, a Fonseca '66 does."

I asked him what his ideal port of memory was, the one he judged everything else by. "A Fonseca '66, of course."

"When did you first drink that Fonseca '66?" I asked.

"When I was 19 years old."

I knew he was not putting me on, and he was not at all a poseur like Borowitz had described, and I did not doubt for a moment that this vintage port—one bottled before he was born, then sipped as a young man, and now remembered half a lifetime later—was absolutely the greatest port he'd ever had and probably would ever have. And also that this meant something very profound to him. At 19, you probably think you'll be tossing back Fonseca '66 your whole life. Only later do you come to realize this is probably not the case.

A few days after the tasting, I logged onto *The Port Forum* to see how the discussion over the 2007 vintage proceeded. Vigorously, I found. Among the biggest questions were, of course, should we buy these? "AHB" from Berkshire, England, wrote: "Do I want any of this vintage? . . . Yes, as I will be in my mid-60s when these are mature and given my genealogy would expect to still be around in the 2030s."

Several forum members expressed excitement that the 2007s might actually be enjoyed young, without waiting for decades, but other port enthusiasts took a dim view of this, as Wiseman had. "Uncle Tom," a commenter from Cambridge, England, suggested that "consumers are not all sold on the idea of drinking VP young, and that anyone reviewing a new vintage should be more focused on the likely performance of the wines when matured, than on their ability to afford immediate gratification."

This response made me stop and think about patience, about denying immediate gratification, about aging gracefully. About wagering money and years in the hope of, perhaps once again, tasting something sublime. Maybe this is the key to understanding vintage ports, maybe the lesson they have to teach. If so, it's no wonder why they're such a hard sell to us here in America.

CHAPTER THIRTEEN

Pouring the Unicorn Wine

AFTER SO MUCH TRAVEL, I was finally home for an extended time. Still, I was restless. My head was buzzing with all the grapes and places and winemakers I'd encountered, those special wines I'd experienced. But I was struggling to make sense of it all. In just a short time, the story of godforsaken grapes had grown much larger than simply Robert Parker versus a new generation of sommeliers and wine enthusiasts.

I started teaching wine classes on Saturday afternoons at a wine bar called Root, in Philadelphia's Fishtown neighborhood. Fishtown is sort of the city's answer to Williamsburg, Brooklyn, with youth and money gentrifying what was once a rough, down-on-its-luck, working-class area. Root's wine list was full of wines from my journeys, such as jacquère from Savoie, scheurebe from Germany, kerner from Alto Adige, and baga from Portugal, and was more adventurous than most restaurants in Philadelphia, or most cities. Root also had some bottles that would be considered rare on just about any but the most cutting-edge wine lists, such as emir from Cappadocia, Turkey, or tintilla from Andalusia, Spain. "We want to have memorable bottles on our list," said Greg Root, the owner. "A pinot noir from Oregon may be delicious, but at the end of the day it's a pinot noir. It's expected. We want wines that are unexpected."

A term had to come into usage, among the wine cognoscenti, to describe the types of rare or obscure or hard-to-find wines that many were after: Unicorn Wine. The idea of the Unicorn Wine was especially in vogue among sommeliers and wine writers in cities like New York and San Francisco, where the term originally referred to certain once-in-a-lifetime vintages from certain legendary, often deceased, winemakers. On the website *Eater*, Levi Dalton, a high-profile sommelier, wine writer, and podcast host, called Unicorn Wines "a new category of wine taking hold in Manhattan—the once in a lifetime bottles that every sommelier dreams of drinking, and bragging about, before they die."

But Unicorn Wine soon came to refer to just about any hard-to-find or little-known wine, inevitably sprouting its own #unicornwine hashtag on social media. For younger wine drinkers, Unicorn Wines began to rival the older generation's Serious Wines. Of course, as with Serious Wine, the definition of Unicorn Wine is ever shifting. What's a Unicorn Wine for me might be different than what it is for a hipster sommelier in Brooklyn or for a young person who's just discovered prosecco and pinot grigio.

So while Unicorn Wine is a slippery term, a bottle of Vara y Pulgar Tintilla on Root's wine list might fit some people's definition (though probably not for some wine geeks). Tintilla de rota is a red grape that grows in the chalky soils around Cádiz, an area known mostly for high-volume whites used to make sherry. In Andalusia, the grape was thought to have gone extinct in the late 19th century due to phylloxera. But a winemaker named Alberto Orte found vines at the turn of the 21st century in an abandoned vineyard and saved it, producing the first 100 percent tintilla to be bottled in almost a century. Tintilla has since been found to be identical to graciano, which is permitted in Rioja blends—but it's still pretty rare.

Yet even at a place like Root, in a gentrifying neighborhood like Fishtown, among affluent thirtysomethings, there was only so far off the beaten wine path that people would venture. "I've found that if obscure grapes are easy to pronounce, they sell well," Root said, pointing to baga and emir—both with four letters and both selling well. Jacquère, on the other hand—the grape that the Wine Mosaic guys in the Alps had told me "isn't that rare"—did not sell well. "Folks couldn't pronounce it," Root said. This consumer behavior seemed to contradict that study conducted by Brock University in Canada, in which wine drinkers overvalued the harder-to-pronounce wine.

My Saturday classes at Root were generally popular, though the ones that sold out were still the classes with familiar topics: Barolo and Barbaresco, Tuscan wines, Spanish wines, Holiday Bubbly. We had a core group, though, who eventually called themselves the Fishtown Wine Club, that came to all the classes. In the beginning, the Fishtown Wine Club knew little more than my university students had, and their tastes ran to old standbys like pinot noir and cabernet sauvignon. But soon enough, we were pouring them Italian whites "beyond pinot grigio" such as timorasso, the nearly extinct wine that had changed my life when I was stuck in Italy, or falanghina, the ancient grape grown in the volcanic soils around Mount Vesuvius, which is believed to be what the Romans called falernian, the most famous wine of antiquity. One of my favorite classes was called "The Pilgrims Didn't Drink Cabernet Franc (But You Can): Wines for Thanksgiving, The Great Pairing Conundrum," in which we poured jacquère, xinomavro, gamay, and the emir from Turkey—whose importer actually promoted the hashtag #drinkturkeywithturkey. By Christmas, the Fishtown Wine Club gathered at Root in ugly sweaters and drained bottles of Loire cabernet franc, baga, and tintilla, where just months ago they would have ordered malbec and pinot grigio. We have a group text chain where we regularly post photos to one another about obscure bottles we find and drink.

One day, I got a text message from Michael McCaulley, the managing partner of the Tria wine cafés who ran the Wine

Safari in the Pennsylvania state store. He wanted me to know that he'd recently acquired a stash of rare Portuguese wines from Colares. Colares is an area near Lisbon consisting of a precious few dozen vineyard acres perched on cliffs above the Atlantic Ocean, near the fairy-tale castle of Sintra. Colares is one of the world's oldest wine regions, boasting some of the few remaining pre-phylloxera vines in Europe that never had to be grafted on to North American rootstock. The reds are produced from a little-known grape called ramisco, which is particularly tannic and acidic, and requires long aging. Though it once was known as "the Bordeaux of Portugal," the remaining vineyards are now in danger of being snatched up by real estate developers eager to build beach homes. Colares could literally cease to exist as a wine region.

So, to recap, Colares represents pretty much everything a modern wine geek seeks out and loves: obscure grape and region, a good backstory, in-your-face tannins and acidity, a wine that's almost extinct. For the cognoscenti, ordering a glass of Colares, then posting a photo with #unicornwine is a no-brainer. But McCaulley has to appeal to more than just those inside the bubble. In other words: Yo, how the hell are you gonna sell Colares in Philly?

That's the kind of wine-sales conundrum that McCaulley has been solving since 2004, when he opened his first Tria wine bar, with more than two dozen wines by the glass on the menu. Now with three locations, Tria is the kind of place where Philadelphians are just as likely to order a glass of listán negro from the Canary Islands or gros manseng from Jurançon in southwest France or refosco from Friuli-Venezia Giulia or a Madeira made from the sercial grape as they are a Napa chardonnay or Australian shiraz. Probably more so. "You get a following," McCaulley told me. "People come to Tria knowing they're going to find something different."

Since Tria is one of my local wine haunts, I hold it up as a prime example of what wine-selling can look like in any city if restaurants, bars, and retailers stopped playing it so safe and embraced new things, and had a little more fun communicating with their customers. The fact that Tria operates in a control state like Pennsylvania, where prices are higher and allocations are smaller, should make their success even more noteworthy to people who sell wine in cities outside of New York and San Francisco. The message: If you can sell geeky wines in Pennsylvania . . . well then, you should be able to sell them anywhere.

"A lot of restaurants make excuses. 'These wines don't work. They don't sell,'" McCaulley said. "I disagree. Customers rise to the level you set for them." For example, McCaulley told me that two of his bestsellers from recent springs and summers happen to be txakolí, the light, effervescent Basque wine made from the hondarrabi zuri grape, as well as Austrian gelber muskateller, which I'd encountered at numerous Viennese heurige. It was the wine I drank with Wolfgang at Weinbau Stadlmann in Maurer, as he reminisced about Tom Selleck's swimming pool. "A lot of these wines are odd to us in the contemporary moment. But they're traditional wines and they're not all that weird," McCaulley said. "We're just rediscovering them. When you explain this to customers, they understand and feel safe ordering. It's not just a fad wine."

The popularity of gelber muskateller, to be fair, would even be a headscratcher in its Austrian homeland. Gelber muskateller literally means "yellow muscat" in German—but for maximum confusion, it's really an old variety known as muscat blanc à petits grains. Every time I told a winemaker in Austria that gelber muskateller sells really well in Philadelphia, they looked at me like I had three heads. But let's remember, it wasn't so long ago when grüner veltliner was strange and new and only the geeks drank it. In fact, if we're being totally honest, putting grüner veltliner on certain lists at certain restaurants and bars would still qualify as adventurous. Root, for instance, has a number of grüner veltliners on its list. But both times we ran a class on Austrian and German wines—called "Don't Fear The Umlaut"—we had to beg people to attend, unlike with any class about Spain or Italy.

Still, gelber muskateller is a good example of an esoteric wine that has found a mainstream following at Tria for over a decade. McCaulley first offered it in 2007, and it's been on the spring menu ever since. "Guests eventually began to request it. They'd say, 'Do you have that gelber in yet?' It's never going to sell like sauvignon blanc, but every spring we've bought at least ten cases. This year we bought 25 cases. And it always sells out without a problem."

McCaulley doesn't expect gelber muskateller or Colares or txakolí to usurp the old standbys. "These wines are not here to replace the usual suspects," he said. "They're here to complement them. There's room for a nice strip steak and there's room for beef tongue." Tria always lists about a dozen mainstream grapes on the menu next to the geeky stuff.

Wine sellers like McCaulley talk a lot about "education," though to be clear their definition of education is not the same as in my university classroom. Their education must end in sales. "We talk about sharing, not selling. It's about starting a conversation between the server and the guest." Beyond that, it takes some extra effort to highlight and promote a wine that people are unfamiliar with, such as Freaky Fridays at Tria, where discounted wines are offered at lunch. On Sundays at Tria, during a promotion called Sunday School, less-known wines are offered at $5 for the first glass. One Sunday, for example, a Gaillac white, a blend of loin de l'oeil and mauzac grapes, was highlighted for $5 on a card clipped to the menu (which included the wine's story, and an explanation that Gaillac is pronounced "guy-yak").

When I think about gimmicks to sell unfamiliar wines, I flash back to one night in Milan when I found myself in a wine bar called La Cieca ("The Blind"), a narrow, easy-to-miss neighborhood joint. It had an amazing list of hard-to-find wines from all over Italy. Beyond that, the real hook at La Cieca was a chalkboard called "Vini alla Cieca," a series of mystery wines, priced from five to nine euros. If you guessed the region and grape of the mystery wine, the wine was free. But there's a catch. All the mystery wines are served in black glasses, making the game extremely challenging and huge fun. It also served a commercial purpose: Even though I guessed two of the lower-priced ones, I ended up dropping twice as many euros than I would have in a typical wine bar in my attempt to guess all five . . . and then more after.

Amid the chatter about education and sharing and storytelling and gimmicks, there's surely one huge question that lurks in the mind of people who sell off-the-beaten-path wines: Can I actually make money on a wine no one's heard of? For McCaulley, offering obscure wines is not only profitable, but a big part of Tria's success. "You don't have to take a low margin on the unknown thing," he said. "Sometimes, the unknown thing has much better value. You make more money and the guest gets better value." Besides, he added: "These wines just make life, and work, more interesting."

The Unicorn Wine business, to be certain, is not all fun and games. It can get rivalrous. If your wine bar is known for having unique wines, customers quickly come to expect to find new bottles that they've never heard of before—and that they can't find anywhere else. Root insists that his distributors sell him exclusive wines that no one else in Philadelphia is pouring. "Our first question is: Do we like it?" he said. "Our second question is: How much does it cost? The next question is: Who else is carrying it?" This can create some tense situations. For instance, Root had fallen in love with the emir on his list, made by Turasan winery in Cappadocia, Turkey, where it's believed winemaking dates back at least 7,000 years. Many believe emir from Cappadocia may, truthfully, be the historic wine of Byzantium. It was selling very well by the glass, and I often enjoyed it.

But a few weeks after Thanksgiving, I happened to be in Tria and saw the same Turasan emir on that list, and made the mistake of mentioning that to Greg Root. By the end of the year, it was off Root's list, never to return.

While I was teaching my Saturday afternoon classes in gentrifying Fishtown, I was living in the suburbs, where the trend of Unicorn Wines had still not arrived and taken hold. What the Fishtown Wine Club was exploring and drinking was far away from the wines that most of my neighbors were experiencing.

One weekend, my friend Rick invited me over to drink some wines from his cellar. Rick, in his mid-40s like me, was my son Wes's Little League coach. He's a great, generous guy who's deeply interested in wine. As a young man in the 1990s, Rick had an older mentor who introduced him to wine and led him down the path of collecting. During that era, Rick collected the Serious Wines that everyone did—Bordeaux, Napa Valley cabs, Super Tuscans—just like Robert Parker told them to.

Rick was getting worried that some of his wines were teetering on the edge of life, since his current home did not have a dedicated wine cellar as his former home did. So we rummaged around in boxes stacked in his basement—which was cool and dark enough for fine storage. We decided to open a 1982 Château Léoville Poyferré (from Saint Julien in Bordeaux), a 1997 Ornellaia (a Super Tuscan from Bolgheri), and a 1989 Dunn Vineyards cabernet sauvignon from Napa Valley. Considering that the current price tags on those wines are, respectively, around $300, $250, and $100, I should mention once again that Rick is a very generous guy. We opened the corks delicately, then let the wines aerate.

Meanwhile, I opened the starter wine that I'd brought, a Rheinhessen riesling—albeit a much better one than I'd had that lonely night at the airport hotel in Mörfelden. This one was a Keller von der Fels, produced by Klaus-Peter Keller, who may be Germany's finest winemaker. Keller makes a bottle that's on my own personal list of elusive Unicorn Wines, his G-Max Riesling, which sells from $800 to $1,400. A magnum of the 2009 was auctioned for over $4,000. I've even visited Keller, sitting with him at his kitchen table, and he didn't pour that one for me.

What I brought Rick was not that one. The Von der Fels 2015 I brought is one of Keller's more standard bottlings, and sells for under $30. But it's still one of the best-value rieslings I know, explosive and stony, with spritzes of wildflowers and lemongrass spice and smoked grapefruit. It's like a brisk walk up a beautiful mountain slope, but leisurely, maybe a blue square, not a black diamond. Even at that price, it's a wine you can age for at least five to seven years.

Even though this Von der Fels riesling is dry (*trocken* in German), Rick's skeptical response before taking a sip was, "Ah, riesling . . . I don't know . . . I always think of riesling as sweet."

His response is sadly familiar to anyone who loves riesling. When we talk about the noble grapes, often we forget that riesling is one of them. It's fallen so far behind pinot noir, chardonnay, and cabernet sauvignon in mainstream prestige, and even further behind merlot and sauvignon blanc in popularity. As a grape, riesling is not rare or obscure, in fact it is one of the more widely planted white wine varieties. But riesling is polarizing and chronically misunderstood. Greg Root says he's mostly given up on carrying riesling by the glass, and instead now offers riesling's offspring, kerner and scheurebe. They sell much better, he says, because they don't have riesling's negative baggage—sweetness.

I actually poured the same dry Keller von der Fels at a private wine class in Philadelphia and received, literally, a revulsive reaction from half of the dozen people attending: "Oh my god, this is too sweet! . . . This is like a cake wine!" The other half loved it, including one who said, "This is like fresh gravel stones, so beautiful!" That tasting party was put on by a member of the Fishtown Wine Club, a lawyer who loves riesling—and because of that has earned the nickname "Sweet Joe."

Yet even before sommeliers talked about Unicorn Wines and before wine bars poured esoteric grapes few people had heard of, the wine cognoscenti was pushing consumers to rediscover riesling. In fact, from 2008 through 2014, each June was marked on the wine calendar as the Summer of Riesling. Examining the ebb and flow of riesling's trendiness gives a good example of just how difficult it can be to educate wine lovers on complex wines.

Summer of Riesling was begun by Paul Grieco, whose revolutionary Terroir wine bar in the East Village influenced the entire wine culture in the mid-to-late 2000s. As Jon Bonné writes in a 2017 profile in *Punch*, Grieco was "the first sommelier punk" at a time when the role of sommelier "had become performance art." Now the young-tattoos-and-wild-facial-hair-hipster sommelier stereotype is a default. But Grieco was the original. Bonné adds: "It isn't too much to say that, for a time, Terroir was America's ur–wine bar. It shifted the way we talk about wine, and inspired dozens of derivative spots along the way." Grieco's crowning achievement was Summer of Riesling, which he started in very punk fashion when he devoted his entire summer wine-by-the-glass list to riesling, basically forcing it on his customers. Summer of Riesling, soon supported by Wines of Germany, spread nationwide. Any bar who was willing to pour three rieslings by the glass on their menu could take part. The riesling menus usually had a punk-inspired design, and temporary RIESLING tattoos were handed out. For several years, nothing was cooler among hipster wine drinkers than riesling.

But then something changed, almost in a snap of the fingers. Right around the end of the last Summer of Riesling in 2014, there seemed to be a sort of weird cultural backlash against riesling. After serious growth in the first decade of the 21st century, riesling sales dropped in both 2013 and 2014, and several of the wine trade magazines suggested that riesling sales were usurped by sickly sweet moscato. Beyond sales, it also seemed that riesling was losing its cultural moment and becoming passé. Haters emerged. On HBO's *Last Week Tonight*, John Oliver offered a cutting putdown, in the middle of a funny skit about safer passwords that are supposedly so preposterous that no one will be able to guess them. Among the word combinations played for laughs: "rieslingdelicious." Not too long after, on HBO's *Girls*, Hannah's mother shouts the devastating line: "Riesling's too sweet, no one likes it!" She was, at that moment, displacing anger at her husband who'd just announced he was gay, but repeated the sentiment again a few minutes later at a dinner party, just to drive home the point. To those selling riesling, this all must have been extremely disconcerting, especially since the "hipness" of riesling was pitched for seven Summers of Riesling to precisely the kinds of people who watched John Oliver and *Girls*. Moreover, if you're a wine person who loves riesling (like me) the phrase "Riesling's too sweet" is especially cringe-worthy. Because we've all heard that over and over—no matter how many ways drinkers are shown otherwise.

So all of this depressed me a great deal. I was a big fan of the Summer of Riesling. But the riesling evangelists in the United States made some grave missteps in their advocacy. Rather than steering drinkers to the drier rieslings, they bullied them toward the sweeter styles, lecturing people with an annual primer on Germany's traditional Prädikat classifications, the complex "superior quality wine" designations based on ripeness levels and must weight: kabinett, spätlese, auslese, beerenauslese. All of them had significant residual sugar; even the lightnest, kabinett, could be pretty sweet. Many sommeliers' educational method seemed to be pouring customers sweet wines, then attempting a sort of *Star Wars* Jedi mind trick: "This is not the sweet wine you're looking for." Or else simply scolding the customer: "Don't you dare say it's sweet!"

On the menus at Terroir, for instance, there was an aggressive sweet-wine evangelism: "The greatest wines in the history of mankind have all been sweet." On riesling, Grieco did not mince words: "Yes, there may be some residual sugar in some German examples. But who cares. In America, we talk dry but we drink sweet. Don't deny it; other than New York City school kids, someone is drinking all that Coke and Dr Pepper and Snapple."

But a problematic truth complicated the evangelists' message. Back in riesling's spiritual home, Germans overwhelmingly prefer dry riesling. More and more wineries have stopped using the traditional Prädikat classifications on their labels. The sweet styles with the confusing nomenclature are in danger of dying off. "Trocken dominates in Germany," Johannes Selbach, winemaker at acclaimed Selbach-Oster in Mosel, told me. "We can show sweeter rieslings until the cows come home. You can talk about them and write about them. But put them on the shelf, and people will buy the dry."

As I've chronicled with grüner veltliner, port, pinot grigio, prosecco, and others, fashions and tastes fly in and out of style. But as Germans moved away from sweet rieslings over the past couple of decades, something strange happened: American sommeliers took up the mantle of the old-fashioned sweet styles. "Americans are keeping the tradition of sweet riesling styles alive," said Christof Cottmann, a marketing manager at Schloss Vollrads in Rheingau. A lament by top San Francisco sommelier Rajat Parr in his 2010 book *Secrets of the Sommeliers* is typical of the American response to the rise of dry riesling. "Unfortunately, young German wine drinkers have almost given up on the off-dry style, preferring instead wines that are so austere and sharp as to be almost without pleasure. Who is going to save the off-dry style? Sommeliers." By *off dry*, Parr means what most normal people call *sweet*.

Near the end of the final Summer of Riesling, the debate over sweet and dry riesling got testy. Terry Theise, who has many German wines in his portfolio, scolded the Germans themselves over the type of riesling they prefer to drink in Germany. In the *New York Times*, Theise expressed his displeasure that sweet styles of riesling were being usurped in Germany by *trocken* riesling—what Theise called "a highly invasive species that wants to swallow up every other style." So frustrating for Theise was the change in taste from sweet to dry that he reached for a rather unfortunate national stereotype: "The omnipresence of dry wines within Germany is a dubious example of this country's temptation to do things in large, implacable blocs."

All of this is why I began to dread the Summer of Riesling as the years rolled on. It began to feel like we Americans were the chumps, drinking the sweet exports that Germans didn't want. When I turned up at one of those hip places with their summer riesling menu, I'd always ask: "Are any of these on the dry side?"

The server or sommelier, in a schoolmarmish tone, would say, "Well, riesling always has a little bit of sweetness."

"Yes, yes," I'd say. "Why don't you just pour me your driest?" Inevitably I would be poured something like a flabby kabinett with lots of residual sugar. At one spot, with one particularly pretentious server, the "dry" option she poured was a 2011 Schloss Vollrads spätlese, with 8 percent alcohol, and with almost 20 grams of residual sugar. I know perceptions can be different, but I defy anyone to tell me that wine isn't sweet. When I tasted that exact same wine at Schloss Vollrads, Cottmann told me: "This is a wine that has absolutely no market in Germany."

The problem, of course, is not that Germans prefer dry, or that American sommeliers want to talk your ear off about the traditional sweet styles. In fact, rieslings from places such as Austria, Alsace, or Australia are predominantly dry. The problem is that the message of evangelists like Theise and Parr and Grieco actually confuses drinkers. I hear from people all the time who tell me that they are interested in riesling, want to like riesling, and yet find themselves staring at the store shelf or the wine list and wondering: Is this one sweet or dry? Experienced wine lovers are not idiots. They can tell a sweet wine when they taste one—and many of them simply don't like it. After too many sweet surprises, they'll just skip riesling altogether. So pour them a dry riesling and save your breath trying to convert them to sweet.

It's no wonder that many American riesling enthusiasts have started simply looking for the word *trocken* on the label. *Trocken* thankfully ensures you'll get something relatively dry, though even trocken can have a few grams of residual sugar. Another way to ascertain dryness is to check the alcohol content. As Selbach told me, "If the alcohol is one digit, it's going to be sweet. If it's double digits, it will be at least dry-ish, or dry."

This seems like a more accurate methodology than those little line graphs you often find on the back of riesling bottles, which give the range from dry to sweet, sometimes with a number scale from one to five, or even one to ten. I've never found these scales useful because they rarely solve the basic issue of "how sweet is this wine?"

A few years ago, the International Riesling Foundation (a nebulous multinational organization if there ever was one) started pushing a universal sweetness scale. Going by this system, a wine that has a 1:1 ratio of sugar to acidity is designated as "dry." That seems pretty convincing, until you realize that a wine with nine grams of residual sugar (with nine grams of acidity to match) would be designated dry. Also, there's a corollary to account for pH. If the ratio is 1:1 but the pH drifts over 3.5, it's now classified "medium dry." For a guy like me, who nearly failed high school chemistry, all this nearly causes my head to explode.

I saw a good example of people's frustration with riesling in the comments section of the wine website *Terroirist*. Under a feature article titled "Searching for America's Greatest Rieslings," a commenter named Bruce wrote:

> *I think the public's problem with Riesling is not that many are "too sweet," it is that for most of us there is no way to look at the bottle and find out which are sweet—some bottles might mention "off dry" which usually means sweet. Most give no clue.*

Bruce was immediately scolded by another commenter, a self-professed riesling expert who told him German wine was "so easy, my 6 YO grandson can pronounce 'hochgewächs'" and suggested he learn not only the term *trocken*, but also *halbtrocken*, *fruchtige*, and *liebliche*. To which Bruce replied:

> *Wish it was as simple as you have found it. This morning I went into Whole Foods (soon to be Amazon) and read the labels on the 10 German Rieslings they stocked. Looking for the four descriptors you mention, I found only one with the term "trocken." None with the other three terms. One had a line graph with the sweetness indicated. The other eight had nothing. That has been typical of my experience over the last 30 years.*

I've come to realize that there's a big difference between a scientific measurement of residual sugar and whether someone perceives that a wine is sweet. I've poured so many rieslings that have little or no residual sugar, only to be told by someone, after their first sip: "This is sweet."

Sometimes, halfheartedly, I say, "Are you sure what you're tasting is sweet? Or is it fruity? There's a difference." Other times, I shrug or chuckle uncomfortably.

Perhaps riesling is always destined to be confusing. That seemed to be the message in the Mosel region, which, with its steep, cool vineyards, is the bastion of the German sweet riesling style. In the tasting parlor of his Dr. Loosen estate, overlooking the Mosel River, Ernst Loosen became very animated as we discussed the sweet/dry divide. "The problem of riesling is that there is not one face! We never get out of this corner of being a sweet wine!" Loosen said. "I hate that it has to be one way or the other. But with wine people, you know it always has to be one way or the other."

A little farther up the river, at Selbach-Oster, Selbach told me, "Sweet is politically incorrect now. We say 'fruity' rather than sweet." Selbach spoke cheerfully and calmly, but clearly the endless debate over sweet versus dry riesling was something that grated. "This feels like a religion. It's like they ask you to swear on the Bible or the Koran. It's black and white, and wine is not black and white."As if to underscore the point, we tasted Selbach's transcendent 2001 Zeltinger Schlossberg spätlese trocken. It was somehow both sweet and dry at the same time. It was like a memory of biting into the freshest, greatest late-summer peach of childhood. This is the kind of riesling that the wine critics are always referring to as "thrilling." "There is a difference between balanced and *sweet*," Selbach said. "Over time the sweetness slowly melts away. After ten, 20 years, the sweetness integrates."

In the rest of Germany, as the climate changes, most winemakers in regions like Rheingau or Nahe or Rheinhessen have no problem getting to alcohol levels of 12 or 13 percent so they can make excellent dry riesling. The winemaking has also dramatically improved. The early dry rieslings in the 1980s were like pure acid, wines that would rip the enamel off your teeth. "We used to call those 'Three-Man Wines.' Because it took two men to hold you down and one to pour it into your mouth," said Cottmann.

But that was decades ago. Now, on my visits to Rheinhessen, I've tasted with a growing number of innovative, younger winemakers who focus on trocken and reject the sweeter Prädikat wines. Many of these producers are among the most sought after in the wine bars of German cities. But several have told me stories about their American importers, such as Theise, who still insist on taking the sweeter wines and sometimes even refuse to taste their dry. While tasting the exquisite wines of Philipp Wittmann, at Weingut Wittmann in Rheinhessen, it's hard to believe those days even existed. "The focus is on the dry here," Wittmann said. "In Rheinhessen today, you see much more focus on dry riesling. We have good acidity, freshness, and a cooling aftertaste, and the alcohol levels are perfect."

One village down the road in Rheinhessen is where Klaus-Peter Keller lives. As the sun set over Rheinhessen, on a cold March afternoon a few months before the final Summer of Riesling, Keller said, with a heavy sigh: "Maybe this is why riesling will always be for the geeks." We were sitting at his kitchen table tasting some beautiful older vintages of Von der Fels trocken, and he sighed because we got talking, of course, about the sweet and the dry. He was clearly tired and so was I. For him, this was the first time home after being abroad, including a trip to New York to spread what he called "the riesling virus."

Keller patiently listened as I aimlessly expressed my preference for dry whites. Then, as if to gently make a point, he brought out a 2007 beerenauslese—one of the sweetest styles of riesling. This one had been made by, of all people, his ten-year-old son, Felix Keller. It was . . . thrilling. Yes, it was a thrilling sweet wine. Klaus-Peter smiled, so proudly, as we tasted.

"In the end, we all believe in the same religion: good wine," he said.

"There are, of course, always religious wars," I said.

He took a sip and sighed again. "Religious wars are bad for everyone."

Back at Rick's house, we finished our glasses of Keller von der Fels. I could tell that Rick wasn't really impressed, but was way too polite to say anything. "It's good. I guess I'm just not much of a white-wine guy," he said.

I understood what Rick was saying. I grew up as a Philadelphia Eagles fan, part of that notorious hardscrabble, bleeding-green fan base. While Eagles fans hate every other team's fans—Cowboys and Giants atop the list—there was a strange form of disdain for fans of the California teams, such as the San Francisco 49ers. From a young age, I was keenly aware that laid-back 49ers fans were to be called "chardonnay sippers," the word *chardonnay* spat like a curse. The implication was clear: Real men did not drink white wine.

Of course, as I grew up and began to prefer wine to PBR, and found myself more at bars with wine lists rather than Jägermeister machines, my enjoyment of white wine improved dramatically. Times, we are told, have changed, and plenty of guys love riesling and albariño and, yes, even chardonnay from places like Chablis or Montrachet. When I teach wine classes, I have plenty of aspirational guys in their 20s and 30s who are game to learn, but there are plenty who recoil at the idea of white wine. "Um, I don't really like white wine," they'll say. Or: "The first duty of wine is to be red," if they're trying to be clever. Or: "I only drink white wine when I'm with my girlfriend." Often, by the end of class, after they've tasted a great white for the first time, they sing a different tune. Case in point is my friend Sweet Joe, who told me that he mostly drank vodka before becoming a charter member of the Fishtown Wine Club.

Yet there is still a significant portion of men who are resistant to white wines. The old-fashioned idea that "real men don't drink white" is sadly still alive and well. I've had friends who work in finance who suggest that it would be career suicide to even broach the idea of ordering a white wine at a business dinner. I have sincerely tried to change this, going as far as to write an article for *Maxim*—the bro-iest of bro magazines—entitled "Real Men Drink Riesling."

Now that we'd decanted two of the older wines, and Rick and I figured the other one was aerated enough, we dove into the evening's main event of tasting them.

While not the types of wines I normally drink, I can say with absolute clarity that the 1997 Ornellaia and the 1989 Dunn Vineyards cabernet sauvignon were both outstanding. In particular, the Dunn Vineyards cab was beautiful, with great structure, lots of fresh vibrant fruit and deep earthy, herbal aromas. In its youth, this wine had been immense and powerful, but now was like an aging athlete, past his prime, whose game was now based on finesse and experience. With 13 percent alcohol and lots of understated savory notes, it was a wonderful reminder that California wasn't always producing high-alcohol fruit bombs. This took me back to the kinds of wine my father drank when I was a younger man.

Then we moved on to the Château Léoville Poyferré, which in Bordeaux's official classification, demanded by Emperor Napoleon III, is considered to be a Deuxièmes Crus, or Second Growth—just below the Premier Cru. The Château Léoville Poyferré's vintage, 1982, is the year that established Robert Parker's worldwide critical reputation. Almost all other wine critics declared 1982 in Bordeaux to be a poor year, too ripe and too low in acidity. But Parker effusively praised the 1982 Bordeaux, especially the bottles from Second, Third, and Fourth Growth producers that, in those years, were still actually good values. He turned out to be 100 percent right. The 1982 vintage in Bordeaux is one of the most sought after and highest priced.

The Château Léoville Poyferré we were tasting had been scored 95 by Parker when he retasted it in 2009, and at that time, he said it would still age another 20 to 25 years. "A brilliant effort," Parker wrote, "stunning concentration, a boatload of power." But nine years earlier, in 2000, Parker had written that the wine was not ready. "Patience is required," he told his readers, promising them "loads of potential."

I am abashed to admit it—after all my Bordeaux-bashing, and all my Robert Parker–bashing—but I loved this wine. It was everything that a wine should be: great freshness, but the years of aging had muted the big, showy fruit flavors and the muscular "mouth-searing" tannins, and now the fruit had become a frame for more savory notes, earthier notes of autumnal forest and wet stones that emerged. This was a deep, dense, profound wine. It was so profound that it actually made me uneasy with my views on Robert Parker. Those doubts started creeping in again. No matter how wrong that I feel Parker is about the "godforsaken grapes" that I love, once upon a time he was totally right about Bordeaux wines like the 1982 Château Léoville Poyferré. Maybe times and tastes have changed, but there was something universal, and ineffable, about that wine.

More significantly, the only way we could have tasted this wine, at this stage in its full maturity, was for someone like Rick (and whoever he bought it from) to store it away in a cellar for 35 years. Had Rick opened it in 2000 or 2009 or any time before now, we may not have been enjoying such a masterpiece. The fact that Parker predicted all this 35 years earlier is quite something.

It's something about wine that's too often overlooked: time. Great wines, Serious Wines, throughout history, were always meant to be aged. Unique grapes from unique places certainly create special wines. But so does long aging. I'll always remember the timorasso with my deconstructed spaghetti carbonara or the himbertscha over raclette, or the chasselas on the Zürich morning, or Moriondo's petit rouge and cornalin in Valle d'Aosta, or the rotgipfler and zierfandler in Gumpoldskirchen, or the schiava at Castel Sallegg. However, the age of Willi Bründlmayer's 1950 sankt laurent—the one we tasted in Langenlois over talk of Dr. Zweigelt and old cars—added an extra layer of Unicorn-ness. Likewise, while I love young mondeuse noire, the age of Michel Grisard's 1989 Domaine Prieuré Saint Christophe, which I drank with the Wine Mosaic team one night at dinner in the Alps, pushed the wine from obscure to Unicorn.

But when I think about Unicorn Wines, I certainly also remember ones that were not necessarily made from obscure grapes. In most of those, age played a key role. As did chance. You don't have to be a sommlier to play the Unicorn Wine game. One evening in 2015, as I rummaged around a New Jersey wine shop's Alsace shelves, I found two extraordinary bottles collecting dust. The first, a 2004 Domaine Zind-Humbrecht "Clos Windsbuhl" Pinot Gris was one of the sexiest wines—and certainly the best gray pinot—I've ever tasted: big, rich, golden, and incredible smoky salty notes, like a campfire-grilled peach dunked into the ocean. The second was 2001 Pierre Sparr

Riesling Schoenenbourg, from one of Alsace's grand cru. It was riesling that had probably been wild and rambunctious with lots of acidity and sweetness in youth. But, after 14 years in the bottle, it had mellowed, with aromas of a football bonfire and notes of honey and apple peel, still clinging to its brightness and snappy, smoky finish. It was like drinking one of those last warm, sunny days of fall. Their yellowing price tags read $23.99 and $20.99, respectively.

I could tell that Rick was especially happy we'd opened his 1982 Château Léoville Poyferré. We sat out on his patio in the spring air and observed how it changed over a couple of hours. I asked him why he'd stopped collecting wine in the late 1990s. He told me that after he got married and had kids and moved into this home in his wife's hometown, he needed to save his money for other, more important things. Lately, he'd considered getting back into collecting. "But wines like this Bordeaux and Napa cabs have gotten so expensive," he said. "I don't know how I'd swing it." Rick is still young and in good shape, and could easily live another 40 years. He'll have drunk through all his 1980s and 1990s Bordeaux and Napa and Super Tuscans. And then what?

I didn't say it that night, but one option would be to collect red wines made from lesser-known grapes like the ones we've been talking about in this book, such as blaufränkisch from Burgenland, mondeuse from Savoie, Montefalco Sagrantino from Umbria, or lagrein from Alto Adige. Just as on McCaulley's wine list at Tria, the off-the-beaten-path bottle can offer much better value.

Or Rick could stock his future cellar with bottles many of my father's generation would scarcely imagine: white wines instead of red. Faced with astronomical prices, maybe younger collectors will move away from the classic reds of Bordeaux, Burgundy, Barolo, and Brunello, and begin to stock their cellars with rieslings, or Loire chenin blanc, or pinot blanc from Alto Adige, or Austrian grüner veltliner. Rieslings, in particular, have long aging potential. For instance, I'll always remember tasting a 1988 Nikolaihof Vom Stein riesling from the Wachau in Austria, which had been aged more than a decade in a gigantic several-hundred-year-old barrel in a winery that dated to the Romans. That wine was like coming upon a magical apricot in a beautiful carved stone in an ancient forest.

Maybe riesling producers need to think of people like Rick. Maybe instead of trying to convert his taste from dry to sweet, they need to show him how elegantly these wines can age. Too often, we pour amazing, single-vineyard expressions of riesling way too young. Once the rieslings age for five or ten or more years, the shocking acidity and small amounts residual sugar begin to balance themselves out. They become utterly distinctive, extraordinary. What you're left with in the bottle is something that can only happen with time. We intuitively know this about great red wines. Few would have considered popping open that 1982 Château Léoville Poyferré in 1987, or 1992, or even (if you listened to Robert Parker) 2002. First of all, the tannins of that wine would have ripped your face off. But after decades, the tannins soften and fall away.

Barolo, like many prestige regions, knows how to best show its wines. By law, Barolo—so tannic in its youth—must age for at least 38 months after the harvest before being released, with at least 18 of those months in an oak barrel. To be considered Barolo *Riserva*, it cannot be released until 62 months, or more than five years, after the harvest. That means that each winery must store bottles and bottles of each vintage in its cellars for up to five years. And even then, we're told to age it another decade or two. Perhaps the top rieslings could be held back in the same fashion.

Riesling offers an unbelievable value compared to Barolo or Burgundy or Bordeaux. Setting aside a cartoonishly unicorny example like Keller's thousand-dollar G-Max, you can find lots of great riesling for under $50 that would age for 20 years, and plenty for around $30 to $35 that would age for a decade.

I don't know if Rick will ever necessarily come around on riesling. But I think that a guy like Sweet Joe, who's more than a decade younger than Rick and me, might someday. After his wine-tasting party, where half of his friends absolutely hated the rieslings, Sweet Joe told me he'd never tasted anything like the 1998 Weingut Ratzenberger spätlese trocken from Bacharach, Germany, that I poured. This was an incredibly confusing wine that was *trocken* (dry) as well as *spätlese* ("late harvest"), meaning it would have been picked late and allowed to ripen fully, with lots of sugar, but that it was technically vinified as dry. Not just a little confusing, but completely confounding to some of Sweet Joe's friends who said, repulsed, "This is like dessert, not a wine!"

"Whoever doesn't like this, don't dump it, pass it down to me," said Sweet Joe. He ended up with six glasses in front of him. Later, he asked me how much that riesling cost, and when I told him $34, he said: "When I become partner, that's the kind of wine I'm going to put in my cellar." Perhaps in 20 years, once Sweet Joe has moved to the suburbs and coaches his kid's Little League team, he'll be generous enough to invite his neighbor over to open it.

CHAPTER FOURTEEN

Looking Forward, Looking Eastward

"I WANT YOU TO TRY SOME JUHFARK," said Tara Hammond, poking around boxes of wine stacked in the East Village apartment that she shares with two roommates. "Wait, where is the juhfark? Ugh, I'm not usually so unorganized." Inside those boxes crowding Hammond's tiny living room may well have been the highest concentration of obscure wines, per square foot, in the United States on that rainy afternoon. We tasted kisi, rkatsiteli, and saperavi from the Republic of Georgia, kadarka and furmint from Hungary, and žilavka from Bosnia. When I'd suggested that the graševina from Croatia and the olaszrizling from Hungary we tasted were also pretty obscure, Hammond waved her hand dismissively, saying, "Oh, graševina and olaszrizling are just different names for welschriesling."

She opened another box. "Ah, here it is! Oh, good, we have juhfark!"

Juhfark is a Hungarian white grape whose name means "sheep's tail" because of the elongated, curved shape of the clusters. This particular juhfark was from a region called Somló, situated on the basalt slopes of what had been, in prehistoric times, an underwater volcano. Wines from Somló have been praised since at least the 11th century, and they were favored by the Habsburg court in Vienna. Somló was considered to be the wedding wine of the monarchy.

As she poured me the juhfark at her kitchen table, I read aloud from the tasting sheet that Hammond handed me:

"Perhaps most well known is the belief that drinking the wine of Somló before copulation would guarantee a boy."

"Copulation?" she said. "I don't think I've ever heard it called that before."

Hammond, in her early 30s, had recently become the New York sales manager for Blue Danube Imports, which focuses on wines from Central and Eastern Europe—essentially the wines of the former Austro-Hungarian Empire. I'd met her through David Foss, whom I'd spent time with in Austria. In addition to Hammond's work with Blue Danube, she was the beverage director at Anfora and its sister restaurant, dell'anima, and was simultaneously studying for her Wine & Spirit Education Trust's diploma, a qualification that is a potential stepping stone to becoming a Master of Wine. This type of hustling and juggling roles is fairly common for an up-and-coming wine pro in New York.

We were supposed to be tasting at Anfora, which has several Blue Danube wines on its list, but Anfora was closed for a private meeting. Plus, Hammond had been delayed because she fell down a flight of steps at her warehouse in Queens, smashing an entire case of wine. She fortunately was unhurt, but when I arrived, she showed me the red wine stain on her yoga pants. "I guess this is why I'm single!" she joked.

Besides trying to establish herself in New York's wine world, Hammond was also attempting to look out over the dashboard, into the blue sky, at what might disrupt the world of wine (choose your own business cliché). There was no doubt that, as I was finishing this book, the wine world was turning its gaze east, toward the Slavic wine regions that once existed behind the Iron Curtain and even farther east, toward the Caucasus. The most cutting-edge lists had wines from places like Hungary, Slovenia, Croatia, Georgia, and Armenia. "I'm doing this to be part of something interesting," Hammond said. She'd become an evangelist for these eastern wines. During the tasting that afternoon in her apartment, it felt like I was witnessing the seeds, the raw material, of a wine revolution.

I'd come up to New York to check out the Whitney Biennial, that every-other-year exhibition of younger and lesser-known artists; I was curious about what was happening or about to happen in the world of contemporary art. I regularly visited the New York wine scene for a similar reason. As cool as Philadelphia may someday become, it's always about a year behind. As nice as the suburbs are, sometimes the future arrives even years later, or maybe never.

After having spent time in Burgenland while in Austria, I was most excited about the wines of Hungary. I'd only been a few miles from the border, and was now feeling stupid for never crossing over. Still, when I was writing my column for the *Washington Post* in the late 2000s, I'd been invited to the Hungarian Embassy by the cultural attaché to taste Hungarian brandy, and then returned a few times to taste the country's finest wines too. So I felt like I had at least a basic understanding.

Hungarian wines, I'd learned, had been among the most coveted in the courts of Europe. This was especially true of the sweet aszú wines of imperial Tokaji, made from furmint, hárslevelü, gelber muskateller, and several other varieties that have been infected by a gray fungus called botrytis or "noble rot," shriveling the grapes nearly to raisins, and concentrating a honeyed nectar. Near the Slovakian border in the shadow of the Carpathian Mountains, Tokaji is the world's oldest wine appellation, predating port by decades and Bordeaux by 120 years. So you could say that Tokaji is pretty much the original Serious Wine. Tokaji wines were served at royal court of Louis XIV at Versailles and Austrian Emperor Franz Joseph sent bottles of

Tokaji to Queen Victoria every year on her birthday—a bottle for each month she had lived. In 1900, on her 81st and final birthday, she received 972 bottles.

Given how little people enjoy sweet wines these days, it made sense that the Hungarian wines we tasted at Hammond's apartment were all dry. The 2012 juhfark we sipped was so opulent and full of volcanic ash, but the acidity kept it light on its toes—unbelievably complex for a wine retailing under $25. It was produced by Fekete Béla, a nonagenarian known as the "grand old man of Somló" who still tends his ten acres by hand. We also tasted Béla's 2011 olaszrizling, a Hungarian weightlifter of a white wine that was herbal, briny, and smoky, almost like a dirty martini. Its retail price: $20. "You cannot find wines like this at these prices," said Hammond, slipping into sales mode.

We moved on to a wine made from kadarka, a thin-skinned red that is believed to have been brought to Hungary by the Turks. This 2015 Eszterbauer Kadarka was precisely the kind of drinkable, spicy, lighter-bodied, "un-American" red that I enjoy. The label read "Sogor" and I asked Hammond if that was the region. "Umm," she said, "it's the region . . . or else it means the love between brothers-in-law?" Actually, as she read from her laptop, *sogor* was Hungarian for "brother-in-law" and was an homage to two who existed in the Eszterbauer family, which has been making wines since 1746. The region, as it turns out, was Szekszárd. We both laughed. These were complicated, unfamiliar wines and what I really appreciated about Hammond was how—unlike a lot of sommeliers—she was willing to admit she didn't remember every last piece of information. In that way, she was just like any of us who love obscure wine and are always trying to learn more.

That afternoon, for instance, is when I learned that kékfrankos is what the Hungarians call blaufränkisch, my old friend Blue Frank. Hammond poured the 2013 Pfneiszl Kékfrankos

"Together Again," made by two sisters, Birgit and Katrin Pfneisl, who were born and raised in Austria, but whose grandparents had fled Sopron, Hungary, to escape Communism. Years after establishing a wine estate in Burgenland, the family eventually reacquired their ancestral Hungarian vineyards, just south of Lake Neuseidl, in 1993. This Hungarian version of blaufränkisch was lighter, less complicated, more easy drinking than ones in Mittleburgenland, but it still had gripping tannins and a distinct herbal note.

It was hard to put one's finger on the flavors of some of the wines we tasted. As we tasted the Croatian graševina (or olasz-rizling, or welschriesling), the flavor was like a strange, tropical fruit and a fragrant, exotic plant.

"It's like a fruit I've never had before," Hammond said.

"Yeah," I said. "It's also like an herb I've never had before."

"As western civilization, we only know what we know," she said. "These are not western wines." I agreed, but I wanted to know what made a wine *eastern*. "Well," she said, "for one, they don't have a lot of the primary fruit we're used to."

This was particularly evident when we moved to the wines from Georgia, which claims to be the birthplace of winemaking, dating back to 6000 BCE. "Producers there find pieces of ancient swords and all kinds of crazy shit in their vineyards," Hammond said. Traditional Georgian winemaking takes place in huge qvevri, clay vessels where wine is fermented and aged, usually buried underground—as I'd learned when I visited Josko Gravner in Friuli. Qvevri wine generally macerates and ferments with the grape skins, and so the whites fall into that trendy orange-wine category. Since Hammond has worked for years at Anfora, she was no stranger to this style of wine before she took the Blue Danube job.

I was poured a qvevri wine from Kakheit, Georgia, made from the rkatsiteli grape. While many Americans have yet to hear about rkatsiteli, it's actually very widespread in the former Soviet states, with more than 100,000 acres planted—and this even after Soviet leader Mikhail Gorbachev ordered millions of vines uprooted during the 1980s. There was something earthy and salty about the rkatsiteli. Even stranger was a qvevri wine made from an extremely unusual grape called kisi, which grows on less than 125 acres. The aroma of the kisi was intense, like a bouquet of pungent flowers in a wedding centerpiece, and in the mouth it was even stranger. This was a white wine with tannins and it felt and tasted like I was licking a terra-cotta flower pot—though not necessarily in a bad way.

Finally, she poured a Georgian red, made from saperavi, which grows on more than 10,000 acres in the former Soviet states. *Saperavi*, which means "dye," is a unique grape because it's one of the few varieties where both the skin and the juice inside have a red pigment. Normally, most red grapes have white juice—and a red wine generally takes its color from the juice macerating along with its red grape skins. This particular saperavi, from a producer called Doqi, was also produced in a qvevri, and while it also had the flower-pot thing and chewy tannins, I loved how rich and how much length it had on the finish. It somehow, for under $20, tasted like wine from antiquity.

In the middle of our tasting, Hammond's roommate popped out of her bedroom to get a glass of water, then disappeared back in. She was launching a feminist publication called *Susie Magazine*. Hammond's other roommate, who was not home, worked as a florist. I asked if her roommates enjoyed all the crazy, esoteric wines that are often available in the apartment. She said no. In fact, her florist roommate was always bringing home bottles of oaky malbec. "I get so annoyed! I tell her, 'One day I'm going to sneak into your room and spread pink carnations all over and see how you like it!'"

All of the wines that Hammond had poured me on that afternoon were unsettling, but in the most positive way. I felt like these grapes, once again, reset my wine compass in some meaningful way. I knew I wasn't the only one, since a growing group of wine geeks were discovering and talking about them, and you were beginning to see them on forward-thinking lists. In fact, it was fascinating to see that the grapes of Central and Eastern Europe, as well as the Caucasus, were even beginning to influence winegrowers at home in the United States.

One domestic bottle that had become of-the-moment was a rkatsiteli produced in New York's Finger Lakes by Dr. Konstantin Frank. I drove up to Keuka Lake to visit Fred Frank, the winemaker and grandson of Konstantin Frank. He told me that rkatsiteli just made sense. His grandfather had migrated to the Finger Lakes from Ukraine, where the grape is prevalent. The elder Frank was always studying cool-climate grapes in search of what might work in the Finger Lakes' microclimate—frigid, but with deep lakes that never freeze, modulating midwinter temperatures, while keeping things cool in the summers.

Instead of taking its cues from Bordeaux and Burgundy—planting cabernet sauvignon, merlot, chardonnay, and pinot noir—the Finger Lakes has always gravitated toward German and Austrian varieties. Finger Lakes riesling is often considered to be North America's finest, and the Franks have also, for years, experimented with other umlauted varieties such as gewürztraminer and grüner veltliner. "What we've learned here over 60 years is that the northern European varieties do best," Frank said. "It's hard not to have a crystal ball, but that's what makes the wine business fun. At least we're not making another Napa chardonnay that's 15 percent alcohol and is so oaky it feels like you're chewing on a two-by-four."

One especially interesting experiment is with rkatsiteli's red Georgian cousin, saperavi. Frank doesn't age his saperavi in a qvevri (rather, 15 months in French oak). It was denser, darker, and fruitier than the clay-vessel version from Georgia—big, ripe blackberries covered in chocolate and spice—but still quite appealing. "These Georgia varieties are not known yet in the United States," he said. "But ten years from now, Georgia could very well become the next Chile." Here's what Frank did not say, but I will: Who knows, but maybe in 20 or 50 years, the Finger Lakes, with its cool-weather Germanic and Slavic grapes, will supplant Napa or Sonoma as the place people turn to for prestige North American wines. As the climate continues to change, it gets less and less implausible.

I was also surprised to find another grape from my journeys growing in the Finger Lakes: Blue Frank. In upstate New York, however, Blue Frank does not go by blaufränkisch or kékfrankos, but rather lemberger, as it's called in Germany. "Even though, yes, there is that stinky cheese connotation," Fred Frank conceded. By his count, at least three-quarters of the Finger Lakes' winemakers have planted some lemberger at this point. Next to cabernet franc, it's become perhaps the region's most important red grape.

Forty minutes east, on Seneca Lake, I visited Fox Run Vineyards, which beyond its acclaimed rieslings produces some of the best lemberger/blaufränkisch/kékfrankos in the Finger Lakes. I met with Fox Run winemaker Peter Bell, a lanky, energetic guy who told me: "We seek linearity here. You could say I'm a skinny guy making skinny wines." He added, "You have to like acidity to like our wines. And that's what Robert Parker doesn't like."

Bell fancied himself a contrarian. When I asked him what made blaufränkisch grow well here in the Finger Lakes, he replied: "Look, I'm a terroir skeptic. There is no terroir here in the Finger Lakes. It's all glacial deposits here. Does it flavor the wines? No fucking way." When I expressed my surprise at that position, he said, "I'm not religious. I'm empirical. I need evidence. I don't operate on faith."

"Wait," I said. "You must have some faith. You've planted a grape called lemberger in upstate New York that you're trying to sell to Americans."

He stared at me for a moment, then smiled. "Touché," he said.

Later, I asked him why he'd chosen to label the grape as lemberger rather than blaufränkisch. "I don't know," he said. "Twenty years ago, when it arrived in the Finger Lakes, that's what they called it. Frankly, I hate both names. But umlauts suck."

Perhaps the notion that grapes from the former Austro-Hungarian Empire or the former Soviet states might someday become the new, widely planted international varieties seems far-fetched. But if you take the long view, it's not so absurd.

One weekend, I was invited by the Smithsonian Associates to an educational event in Washington, DC, called "The Origins of Wine Civilization: From Ancient Vines To Modern Expressions." Among the presenters at this event would be my ampelographic hero, José Vouillamoz. I had not seen Vouillamoz since our raclette dinner in Valais, Switzerland and I was excited to to reconnect with him.

I met Vouillamoz the evening before the event, at a welcome reception thrown at the Georgian Embassy, where many bottles of the country's wines were poured. Vouillamoz talked about the first time he'd been to Georgia, in 2003, when he met winemakers who were so isolated they'd literally never even heard of cabernet sauvignon, pinot noir, or chardonnay. After the reception, he and I ended up at an "international street food" spot on 14th Street called Compass Rose. Amid the typical mélange of lamb kefta, poké, chicken skewers, and Asian steamed buns were two Georgian dishes that we ordered: a bean stew called *lobio* and a traditional hot cheese-filled bread called *khachapuri*, into which the server stirred a fresh-cracked egg. The house wines at Compass Rose were Georgian saperavi and a Georgian white made from a blend of rkatsiteli and a grape called mtsvane. In fact, the menu of about 20 wines listed nine from Georgia, including four orange wines (listed under "Amber Wine"). For a moment, I had a weird *Twilight Zone* experience where I thought: Have I entered a parallel universe where all the French and Italian things we know have been replaced by things from the Republic of Georgia? These days obscurity is less obscure than it's ever been.

We ordered a bottle of žilavka, a white wine from Bosnia, which was poured by a bartender with a nose ring who told us she was an artist. Vouillamoz had arrived directly from a wine conference in Lisbon and he talked excitedly about the potential of the Portuguese grapes. He had to fly back to Switzerland the day after his Smithsonian presentation. I asked about the late Jacques Granges's mountaintop "vines in the sky" at Domaine de Beudon, and who was tending to them after his death. A group of neighboring producers, he said, was volunteering to help Granges's wife with the work this year. "I don't know what happens next," he said. Then he told me the disconcerting news that Giulio Moriondo, in Valle d'Aosta, had recently leased his "Mother Vines"—the ones dating to 1906 that produced the sad and beautiful Souches Mères I'll always remember—to another winemaker. Moriondo would no longer make Souches Mères.

The Bosnian žilavka we drank at Compass Rose, unfortunately, was not so memorable. It tasted a little bit like a flower pot full of cigarette butts. After finishing our first glasses, Vouillamoz asked me if I liked the žilavka. I kind of smiled and shrugged, and he did the same. Neither of us wanted to admit we didn't like it. It was the same feeling I'd had in the Alps when my father had texted me, "Are the wines any good?" I still felt guilty every time I encountered an enigmatic, little-known grape that I didn't like, as if I'd failed a stranger who needed a friend.

"I think I'm going to get another bottle," Vouillamoz said, ordering a young blaufränkisch from Judith Beck, a biodynamic producer in Burgenland, which we both agreed was fantastic: earthy, fresh and sporty, drinkable, with a little funky, spicy edge.

However, the artist-bartender with the nose ring seemed concerned that we weren't finishing our žilavka. We assured her all was fine, that it just wasn't our thing, and suggested she pour it for someone else who might appreciate it more. She told us that the Judith Beck blaufränkisch was new and she hadn't had a chance to taste it yet. We offered her a little splash. "It's very light," she said, swirling and sipping. It was difficult to tell whether "very light" was a good thing or not.

The next morning, the "Origins of Wine Civilization" program was held in the basement of the Smithsonian's copper-domed S. Dillon Ripley Center. The crowd of several dozen Smithsonian Associates members ranged from older middle-aged folks to young senior citizens. Our host, David Furer, who writes a column for *Sommelier Journal*, was a sort of caricature of that old-time American oenophile, casually mentioning that he'd lived in London's Soho for years, and referring to wine as "this hedonic beverage with its myriad of expressions." All day long, between lectures, we would be tasting wines from the world's oldest wine cultures: Georgia, Armenia, Turkey, Israel, Palestine, and Lebanon.

The first speaker was Patrick McGovern, well known in the drinks world as an archaeologist of ancient beverages, and the author of several books including *Ancient Wine: The Search for the Origins of Viniculture*. McGovern pioneered the field of biomolecular archaeology and his lab at the University of Pennsylvania discovered the earliest alcoholic beverage in the world, dating to 7000 BCE in China. He has also consulted with Dogfish Head brewery on its Ancient Ales series. He is sometimes called, unsurprisingly, the Indiana Jones of Ancient Ales, Wines, and Extreme Beverages. (Note to Hollywood: maybe archaeology is in need of a new iconic swashbuckling hero?)

McGovern showed us slides of wine jars from Iran that dated to 3500 BCE and a painted Egyptian amphora dating to 3000 BCE that depicts people picking grapes from trellised vines. "Almost all animals are attracted to sugar and alcohol," McGovern said. "But humans are particularly suited for fermented beverages." He suggested that the earliest humans likely consumed a primitive form of wild, fermented fruit. As Roland Velich had told me back in Burgenland over blaufränkisch: Wine is older than art.

But the big question for researchers seems to be: When and where was the first grapevine domesticated? The Georgians, who had hosted the welcome reception, were keen to say it happened in Georgia around 6000 BCE, and they base their claims on a handful of fossilized grape seeds that were found in ancient qvevri. McGovern, however delicately, disagreed. He told us the oldest grape seeds found have been carbon-dated to only 2000 BCE and the oldest Georgian qvevri ever found date to around 800 BCE—pretty old, "but a long way from 6000 BCE." McGovern said he leans instead toward Anatolia in southeastern Turkey as the likely birthplace of wine, between the Tigris and Euphrates rivers, in the so-called "fertile crescent" where agriculture first developed around 7000 to 9000 BCE. Jars with the residue of ancient wine have been found there dating back to the Neolithic age.

Despite all the things we don't know about the origins of wine, one thing is clear: As soon as the earliest civilizations made and drank the first wine, it was followed soon after by the first wine critics and gatekeepers, who judged, ranked, and classified its tastes. McGovern showed us wine vessels from ancient Eygpt that were stamped with the harvest year, the location along the Nile where it was produced, and the name of the pharaoh who owned the vineyard. "Those were the first wine labels," he said. Unsurprisingly, wines from certain vintages, places, and pharaohs commanded more prestige than others.

Following McGovern's lecture, we were poured wines from Armenia, Georgia, and Turkey. From Armenia we tasted a white called voskehat, with piercing acidity and an odd incense aroma, and a red called sereni that was savory, spicy, and cedary. Then we tasted a red wine made from areni, full of attractive Mediterranean spices. I'd actually seen a handful of areni wines from Armenia on extremely cutting-edge wine lists.

As it turned out, the areni was made by a partnership that involved Paul Hobbs, a jet-setting California winemaker who worked with Robert Mondavi, Opus One, and Simi before starting his own winery. Hobbs now makes wine in countries all over the world, and it's worth noting which ones and with which grapes. In 1988, he saw the potential of malbec as the signature grape of Argentina, and is widely credited with helping to ignite malbec's popularity among American wine drinkers. In 2014, Hobbs announced a partnership with Mosel riesling producer Selbach-Oster to plant vineyards in New York's Finger Lakes. And since 2008, Hobbs has been working with areni in Armenia. He actually joined our tasting via video chat from his home in California. "I did a lot of work developing malbec in Argentina," he told us. "This has been even more difficult. But areni may actually be the best old-world variety."

From Georgia, we tasted classic examples of qvevri wines made from rkatsiteli and saperavi. Lisa Granik, a Master of Wine from New York, asked us not to call them orange wines. "I plead with you to call them *amber*," she said, adding that many consumers hear "orange wine" and believe it to be a wine made with oranges. Granik tried to explain the strange, unfamiliar taste of qvevri wines. She suggested that with qvevri, fresh fruit flavors became more like dried or cooked fruit, though the eastern flavors are more complicated than that. "With the amber wines, it's like 'What's this party in my mouth?'" she said. "I ask people to remember the first time they tasted wine at all. It was a strange experience, right?"

After Georgia, we were poured wine from Turkey made from grapes like narince, öküzgözü, and boğazkere. If I'm being totally honest, by the time the Turkish wines got poured, even I was suffering a little bit from obscure-grape fatigue. But then I tasted the boğazkere. It was deep, dark, and tannic, with eucalyptus and licorice, and it reminded me of the sagrantino I tasted back in Montefalco with the black truffles.

Daniel O'Donnell, the Californian who made this Turkish wine, stood up to talk, in shorts and a T-shirt. "As a California winemaker," he said, "I want to apologize for what we did to chardonnay in the 1990s." O'Donnell, like Hobbs, was now making wine in Asia Minor, for a company called Kayra Wines. "Making wine in Turkey, I've come to understand that the world doesn't need another shitty merlot," he said. "A wine's job, first and foremost, is to be interesting. I prefer to spend my time in Turkey with these exciting varieties."

Finally, after lunch, it was José Vouillamoz's turn to present. He told us that scientists have absolutely no idea how many grape varieties exist in the world. Including table grapes and raisin grapes, there may be as many as 10,000. China has more species than anywhere else, most of them wild species that were never domesticated. In one slide, Vouillamoz showed how the root word of *wine* predates the advent of Indo-European languages, and is the same in nearly all language families.

All we really know for sure is that 1,368 grape varieties are used to make wine. "And 99.9 percent of the world's wines are made with the *vitis vinifera* species," he said. "That .1 is because of you guys, you Americans, who insist on making wines with non-vinifera grapes like *vitis labrusca* and *vitis riparia*." He was talking about grapes like concord, which I had eaten with my son Wes at the beginning of this journey. Vouillamoz chuckled and said, "I've tasted some of the wine made with *vitis labrusca* and *vitis riparia*. Though I would not call this wine. I would call it a liquid bonbon."

Then Vouillamoz regaled us with the story of zinfandel. For decades, many Californians argued that zinfandel was indigenous to their state. But since no *vitis vinifera* was ever indigenous to the Americas, this was completely false. We now know that zinfandel arrived in the United States in the 1820s, delivered from the Imperial Habsburg nursery in Vienna, wildly mislabeled as "Black Zinfardel [sic] of Hungary." Years later, it found a match in a grape called primitivo in Puglia, Italy. But where did primitivo come from? Researchers wanted to know. The search became known as "Zinquest." In 2001, Vouillamoz was in the lab, running hundreds of DNA tests, on the day that zinfandel's DNA finally found a match: a rare grape from Croatia called tribidrag, also known as crljenak kaštelanski. "We definitely popped open some champagne that day in the lab," he said.

When one thinks about the amazing breadth of wine grape diversity, Vouillamoz said, it's almost like considering the many breeds of dogs. To carry out this analogy: Finding out where and when the grapevine was first domesticated is like trying to find the moment in history when the wolf was domesticated and became a dog. Vouillamoz jokingly asked us to consider the plight of the poor domesticated grapevine. "We torture this vine," he said. "We cut its hair, its legs, and its arms. The vine is stressed and puts out more fruit in an attempt to reproduce. It's torture!"

When I thought of tortured grapevines, I immediately flashed to a vineyard and winery that I visited in Bethel, Vermont. It was early December, and snow blanketed the ground. White-knuckled, I had driven up steep, icy Fort Defiance Road and then up Mount Hunger Road to get to La Garagista, to visit Deirdre Heekin and Caleb Barber, who are bravely making wine in the frigid Green Mountain State.

Mountains and borders breed a certain kind of quirkiness, as I saw throughout my wine journeys, in places like Valle d'Aosta, Alto Adige, and Friuli. But the first place I learned this was during my college years here in the Republic of Vermont. Here, we voted neither Democrat nor Republican, but rather Progressive, as Bernie Sanders was first elected to Congress during my senior year. We could watch Montreal television stations in French. In deep Lake Champlain there was said to live a mythical, prehistoric "lake monster" as well as a real-life plague of invasive sea lamprey. When, just a few years ago, the state of Vermont decriminalized marijuana, I was mildly surprised to realize it had actually been criminal at one time. As far as fermented beverages go, Vermont likely boasts the most kombucha-on-tap sold, per capita, in the world. Or else, it's better known for cult craft beers, where beer geeks make pilgrimages to brewers like Hill Farmstead or The Alchemist, where people drive hours to stand in line for cans of Heady Topper. The first vineyard in Vermont was planted in only 1996.

The vineyards at La Garagista are full of hybrid grapes, crosses of European *vitis vinifera* with native North American species. Hybrids are part of the .1 percent, non-vinifera wines that Vouillamoz had mocked at the Smithsonian. Vermont doesn't have the same favorable microclimate as the Finger Lakes, and the only vines that would survive midwinter temperatures of minus-20 were hybrids. "We tried to grow riesling and blaufränkisch, but they just wouldn't come to fruit," Heekin said. "So we pulled them out and planted the hybrids. We knew the hybrids would grow here. They're from a uniquely American melting pot and they're uniquely suited for here. In Vermont, what we do have are some of the oldest soils on the planet."

Some of La Garagista's hybrid grapes, like marquette, la crescent, and frontenac, were bred in a lab at the University of Minnesota's Horticultural Research Center in the 1980s. These grapes have complex parentage; if they were dogs they would be mutts or mongrels. La crescent is a cross between muscat and an unidentified native grape from the *vitis riparia* species. Frontenac comes from a grapevine that grew wild in Minnesota that was crossed with at least eight different grape species. Marquette's grandparent is pinot noir, but it is made up of several non-vinifera species such as *vitis riparia*, *vitis labrusca*, *vitis rupestris*, and others. "Marquette is like pinot noir in the field," Barber said. "It's very sensitive. It's problematic, it's tedious. You've got to watch it all the time."

Hybrid grapes have a bad reputation in the wine world. For the most part they are banned for winemaking in Europe. Baco noir, created during the phylloxera epidemic in France, used in Gascony to make the wine that's distilled into Armagnac brandy, is one of the few hybrid grapes that's permitted in an official European appellation.

In the United States, however, there's a long tradition of making wine with non-vinifera grapes like norton (largely from the species *vitis aestivalis*, the "pigeon grape") in the Midwest or muscadine (species *vitis rotundifolia*), which has been cultivated since the 16th century in the American South, and catawba (a cross of *vitis labrusca* and *vitis vinifera*) that was used to make a popular sparking wine in the mid–19th century. A journalist visiting from London in the 1850s claimed the catawba sparkling wine from the Ohio River Valley rivaled that of the German Rhine and "transcends the Champagnes of France." After the Civil War, much of the catawba was pulled out or abandoned in favor of concord.

I once visited a historic orchard at the preserved Filoli estate in Woodside, California, which preserves more than 200 native and hybrid grape varieties with names like Cascade, Goff, Kendaia, Lucile, Brant, Ruby, Clinton, Gaertner, Yalhalla, and Delicatessen. A lot of Americans might find it easier to order a Ruby, a Lucile, or a Delicatessen than, say, blaufränkisch, olaszrizling, or rkatsiteli. But as of now, as far as I know, no adventurous winemaker has undertaken to make wines from any of them.

One major problem for making wines with hybrids is the levels of acidity—they either have very little, or they have so much searing acidity they'd strip the enamel from your teeth. Heekin said that Cornell University and others have been experimenting with native northeastern yeasts. She firmly believes that the answer to taming some of the hybrids' acidity will come from a better understanding of how to use these native yeasts, rather than commercially produced yeasts. "Hybrids want to be dealt with honestly," Heekin says. La Garagista is firmly in the camp of natural winemaking, with organic farming techniques, handpicking and foot-crushing with very little intervention, and no oak. "We hope some day to have an anfora," says their website.

Heekin and I tasted wines from plastic flextanks in the barn that doubles as her winery, as she shoveled logs into a wood-burning stove. They don't claim to make world-class wines—yet.

And they produce only about 3,000 bottles per year. La Garagista's first vintage was released in 2010. Until 2016, Heekin and Barber ran a restaurant called Osteria Pane e Salute in Woodstock for 20 years, but closed it to focus on winemaking full time. "We are still literally at the beginning of terroir here," she said. They're still figuring out the winemaking, but what excites me about La Garagista is that, by working with the hybrid grapes, they're actively adding new letters to wine's alphabet, new colors to wine's paintbox. Vermont is not the fertile crescent between the Tigris and Euphrates, nor the Republic of Georgia, nor Bordeaux, nor Napa Valley, nor even New Jersey. But perhaps these hybrids represent the humble origins of a new epoch in winemaking.

We sipped a strange white from a grape called Brianna (named after horticulturalist Elmer Swenson's granddaughter in Nebraska), which was full of strange juicy fruit, lemon curd, and cantaloupe—but it seemed to track *backward* on the palette, starting with what would typically be the finish. Heekin and Barber use Brianna to produce a *pétillant naturel*, an old-time, rustic-style wine made *methode naturale*, meaning the juice is put into the bottle before it's finished fermenting, creating a lightly sparkling wine. *Pétillant naturel* has become super-popular with natural wine enthusiasts the world over, who call it by the nickname *pet-nat*. La Garagista's pet-nat is called Ci Confonde—"It confuses us" in Italian. As a pet-nat, Brianna was very pleasant, though with stiletto-like acidity. It's the kind of wine that might appeal to a drinker who enjoyed cider or sour ales, and Ci Confonde is even topped with a bottle cap.

Over some sliced meats and cheese, Barber popped off the bottle cap of another Ci Confonde, this one a red pet-nat, made from 100 percent marquette grapes, the sensitive, finicky hybrid. As the fizzy purple wine tumbled into the glass, a thick pink foam formed in the glass. I took one sniff of the earthy, fertile, pungent nose, and it took me all the way back to Pieve San Giacomo, to 19-year-old me in my Phish T-shirt, sipping that red, frizzante Gutturnio that Paolo would pour for me all those years ago. That cold, snowy night in Vermont, I felt like maybe I had come almost full circle.

CHAPTER FIFTEEN

How Big Is Your Pigeon Tower?

"THE BLACK MAUZAC ISN'T REALLY A MAUZAC, if you know what I mean," said Florent Plageoles, as we tasted from the stainless steel tanks at Domaine Plageoles in Gaillac.

Here's the thing that was beginning to worry me: I knew *exactly* what Plageoles meant. Mauzac blanc is used in southwest France mostly to make crisp sparkling wines with a delicious apple-peel aroma. In fact, wine historians believe that sparkling wines made with mauzac in southwest France predate Champagne by at least a century, making them the world's oldest bubbly. Mauzac blanc is related to mauzac vert, mauzac roux, and mauzac rosé. Mauzac noir, on the other hand, is completely unrelated to the other mauzacs, and is instead genetically close to fer servadou, which is actually called braucol in Gaillac.

Yes, after so many years wandering down the wine labyrinth, I had come to understand so much of this grape madness. I had resigned myself to the fate that there would always be a different, little-known region to experience, another forgotten variety to taste, a new alias to memorize. Yet my visit to southwest France, the region that Bordeaux had tried to crush 500 years ago with its *police des vins* code, felt particularly important.

With Plageoles, I tasted a half-dozen different mauzac expressions from the tank, and then the braucol, and then in his tasting room he poured a white from the ancient grape ondenc, which tasted like a strange, bitter pear. Then we tasted a red made from duras, which takes its name from the French word for "hard," because of its hard rootstock, and how difficult it can be to grow. The duras was full of fresh acidity and black pepper, just like a good Carnuntum zweigelt. "This only grows in Gaillac," Plageoles said of the duras. "Nowhere else in France. Nowhere else in the world, actually." There is roughly as much duras in the world as there is petite sirah. Finally, Plageoles poured a red wine made of 100 percent prunelart (sometimes spelled prunelard). Prunelart was nearly extinct by the end of the 20th century, with less than five acres in 1998. Now there are a few dozen acres more, but it wasn't even officially allowed in the Gaillac appellation until the spring of 2017. Despite being full-bodied, prunelart was also delicate and floral—I don't know that I've had a wine that smelled so much like fresh violets.

Gaillac wasn't the only appellation that it felt important to visit. A few days earlier, I'd gone to the village of Fronton, a half hour north of Toulouse between the Garonne and Tarn rivers, to taste wine made from négrette, another of the godforsaken grapes. Négrette remains one of the most mysterious wine grapes in the world—even Vouillamoz, Harding, and Robinson, in *Wine Grapes*, have no answers as to its origins. The best guess remains that it was brought to France from Cyprus in the 12th century by the knights of Saint-Jean of Jerusalem, returning from the Crusades—one myth that has yet to be disproven. Oddly enough, négrette was once a staple in California, where until 1997 it was known as pinot St. George. The Inglenook wines that were considered the finest in Napa Valley in the 1940s and 1950s usually had some pinot St. George, or négrette, in the blend. I don't know what has happened to the pinot St. George now that Francis Ford Coppola owns Inglenook.

In the Fronton appellation, négrette must make up at least half of all red wines, often blended with cabernet franc, syrah, or fer servadou. At Château Bellevue La Forêt, its vineyards in a former royal forest, they bottle their 100 percent négrette—dark and redolent of grilled fruit and smoked herbs—labeled as

Mavro, Greek for "black," a nod to the grape's mythical origins. However, so far no definitive, scientific link has been made to the mavro grape, which grows on Cyprus.

While in Fronton, I met a rugby player named Nicolas Roumagnac who owned Domaine Roumagnac. Roumagnac looked like a grown-up, dark-haired version of Tintin and, on the day I visited, he had a black eye from a recent match. Roumagnac told me his négrette is the official wine of the Toulouse professional rugby club, which is considered one of the best in the world. "My story here is not folkloric," he said. He grew up nearby and partnered in 2009 with a local winemaker to create his label. "I'm of a new generation and I wanted to give a new spirit to négrette."

Roumagnac believes in blending négrette, since it is so distinctively perfumed. His "Authentique Rouge" bottling was a blend of mostly négrette, along with syrah, cabernet franc, and even some cabernet sauvignon. The result is a rugby captain of a wine, ripe blackberries and spice, with an incense note that was somehow, intangibly Middle Eastern. Roumagnac called the wine "gourmand," but it was attractive, like that slightly paunchy guy who you find out is an amazing dancer. It's hard to imagine that even the most committed Serious Wine people would not like Roumagnac's "Rendezvous" bottling, a black-and-blue blend of négrette and syrah that was dark and fresh, with flashes of tobacco and coffee, and laced with licorice. "We are alone in the world to vinify négrette here in Fronton," he said. "We are very strange. Our négrette is very strange."

Yet equally strange as Fronton and its négrette were the red clay soils of Marcillac, about two hours northeast of Toulouse. So distinctive is Marcillac's *rougier* earth that as I drove through the village of Clairvaux-d'Aveyron, every building I passed had a reddish hue, clearly made from local stone. At Domaine du Cros, I met Philippe Teulier, his son Julien, and their dog,

Mansois—which is the local name of fer servadou, or braucol, in Marcillac. *Fer* means "iron" in French, or it comes from *ferus*, Latin for "wild, savage"—in either case it suggests how difficult the grapevine of fer servadou/braucol/mansois is to work with.

For years, I have sought out Domaine du Cros's classic, everyday mansois-based wine called Lo Sang del Pais, or The Blood of the Country—I love its savory mix of grilled peppers and ripe cherries spiked with anise, its wildlflower aromas, and its wild, ferrous core that's like fresh blood. It may be one of most drinkable wines for under $15 you'll ever find, light-bodied and low alcohol, which is good because you'll want to buy two bottles. I asked why, in the United States, Lo Sang del Pais is labeled by the synonym fer servadou and not mansois. Philippe Teulier shrugged. "It's a mistake," he said. "Here, we call it mansois."

Over veiny, blue Roquefort cheese made from local Laucane sheep, the Teuliers poured me older expressions, including a 2009 Vieilles Vignes, made from 50- to 100-year-old vines. As mansois ages, the wildflowers mellow into rose, the pepper tames to exotic spice, the anise segues into espresso coffee, and the cherries become dipped in dark chocolate; yet the wine never loses that blood-iron backbone. I told the Teuliers that even with age, to me the wine was still incredibly fresh and drinkable. With Julien translating, Philippe said: "Yes, drinkable. It's important to say that, to make that clear. Drinkable is a high compliment, thank you."

After Marcillac, I drove west to Cahors, a medieval town dramatically sited on a peninsula in the River Lot, with an imposing 14th-century bridge. Cahors is the origin of malbec, which most people now know as a wine from Argentina. Here, however, malbec has always been called cot—known since the Middle Ages as the robust and brawny "black wine" of southwest France. Cot from Cahors, by most measures, is an obscure wine to the modern drinker.

On the other hand, malbec from Argentina—softer, fruitier and more approachable—is a massive mainstream success. "Certain wines are so popular it's practically impossible to remember when they were not," wrote Lettie Teague about malbec in the *Wall Street Journal* several years ago, adding: "It was a wine discovered by regular people—not sommeliers." There are now more than 65,000 acres of malbec growing in Argentina, compared to only around 15,000 acres in Cahors. So popular is Argentine malbec that even producers in Cahors have also started labeling the grape as malbec instead of cot.

The producers in Cahors seemed pretty defensive about Argentina's rendition of their native fruit. "Malbec from Argentina is different from here," said Patricia Delpech at Château du Cèdre. "In Argentina, they make wine without much acidity."

"There's good Argentine malbec, we've tasted it," said Daniel Fournié at Château Haut-Monplaisir. "But we think it's a small proportion of what's made."

I visited a young natural-wine producer named Simon Busser, who farms using horses instead of tractors. Busser, ponytail and beard, wearing a Mexican poncho, was hand-rolling a cigarette as I pulled up to his home. As we toured his winery, he pointed out a trailer on his property. "My friend asks if he can bring his trailer and live here a while. Now, he's been here five years!" Busser inherited the vineyards from his father more than a decade ago. "Ah," he said, "my father used a lot of chemicals here. I didn't want to make wine like that. It took me ten years to change the soil."

"Cahors wine doesn't have a good reputation," Busser said. "People in France think cot is too rustic, very strong with lots of tannins." He insists on calling the grape cot, rather than malbec. Instead of long-aging in barrels, he ages his malbec partly in concrete tanks, ferments with native yeasts, and limits his sulfites, all the processes natural-wine lovers demand. But

Busser's wines are always rejected by the official appellation, Cahors AOC. "They say my wine isn't typical. It's too acidic." He must bottle his wine as *vin de table*, France's lowest designation. It doesn't really matter. More than half of it is exported, to the United States and elsewhere, where Busser's Pur Cot is a darling of natural-wine enthusiasts. Busser's cot was like no malbec I'd ever tasted. This was hippie malbec, refreshing, with free-love acidity, and wild-haired minerality.

I ate a lunch of homemade foie gras and carrot salad with Busser, his wife Miriam, and their elementary-school-aged daughter, whom they home school. "She learns on her own," Miriam said. "When she wants to learn something and she's ready, I teach her." It seemed like an apt metaphor for the natural-wine approach to farming and winemaking.

Even though I loved every stop of my whirlwind tour of southwest France's wine regions, the place that won me over most was Gaillac. It fed my passion for all things obscure and quirky. For instance, a major architectural attraction of the Tarn landscape, promoted by the tourism office, was "La route des pigeonniers," which highlighted the region's 1,700 pigeon towers "to satisfy the curiosity about these 'houses for our feathered friends.' " Pigeons were kept for their guano, or pigeon shit, which in the 11th century had been decreed by the monks as the only fertilizer allowed in Gaillac. Young pigeons were sold as a luxury food item.

"The *pigeonnier* was a sign of wealth. The size of your pigeon tower showed how rich you were," said Alain Cazottes, winemaker at Domaine des Terrisses. Cazottes and I were at lunch, and I was eating young pigeon. I was certainly amused by the idea of a pigeon tower being some sort of status symbol, but it's a historical fact. From the 16th through the 18th centuries, rivalrous land owners built their *pigeonniers* ever taller, ever more ornate, and ever more expensive. It sounds comically insane. But when I think about it, I guess keeping a fancy tower full of potentially lucrative pigeon shit is no less crazy than dropping thousands of dollars for a Serious Wine to show off one's status.

The young pigeon I was eating paired deliciously well with Cazottes's Domaine des Terrisses white blend made from loin de l'oeil and mauzac. Loin de l'oeil is a complete mystery to ampelographers. Perhaps it came from a wild local grapevine? Or perhaps, as one winemaker suggested, "It's magic. Maybe it came from the sky?" The pigeon also paired really well with Domaine des Terrisses blend of braucol, duras, and syrah.

After lunch we took a walk through Cazottes's vineyards. As we did, the sky began to darken. "All the bad weather comes from that direction," he said, waving to the northwest. "That's Bordeaux. All the bad weather and diseases come from over there." This was in contrast to warm, dry *vent d'autan* wind, blowing from the southeast during autumn, that the strange grapes of Gaillac love.

We stepped into Cazottes's winery, where as a seven-year-old he'd fallen into a vat during fermentation; he would have died if his father had not pulled him out by his leg. "I've been working in the winery ever since." We tasted braucol and prunelart straight from the barrel. "Everyone in Gaillac knows that prunelart has big potential," he said. Cazottes told me that old-timers know prunelart was once surreptitiously, and illegally, added as a "medicine wine" to Bordeaux wines to fix the color, aroma, and tannins during difficult vintages.

Later that night, I had dinner with Nicolas Hirissou, in his farmhouse at Domaine du Moulin. Hirissou was a tall, jokey goofball, who'd played team handball at a high level, and who had apprenticed in Napa Valley in 2001. He was excited to talk about America. "I love America and I like Americans. I want to sell as much wine there as I possibly can."

I asked him what he learned about wine in Napa. "About wine? Nothing. But I learned quite a lot about American people."

Dogs and cats ran around Hirissou's baby Coline, and his six-year-old son, Matteo, fed logs into the wood stove. We popped open Hirissou's sparkling mauzac, simple but refreshing, like drinking a perfect golden apple. We moved on to his white blend of loin de l'oeil and sauvignon blanc, which was fleshy and balanced with apricot and grapefruit—something that a lover of grüner veltliner, or even maybe a chardonnay drinker could be convinced to try. Then, we switched to reds, first to his entry-level *rouge* blend of duras and syrah, full of berries and spice and dangerously drinkable. You can buy the Domaine du Moulin Rouge in the United States for around $13. Like Domaine du Cros's Lo Sang del Pais, it's the kind of wine with which you might convince even your cheapest, most conservative wine-drinking friend to be adventurous and live a little.

Finally, we did a vertical tasting of Domaine du Moulin's best, reserve braucol, called Florentin, tasting back to 2007. Earlier, he'd showed me the Florentin vineyard, which was full of gravel and stone. "Braucol is very difficult," Hirissou said. "If you don't have maturity, it can taste like vegetables." For that reason, it must ripen late, which is always a gamble.

I admired Hirissou's Florentin, which was his ambitious attempt at showing braucol could make a Serious Wine. They were dark, rich, and intense. "Here we like color. If it's not dark, it's not red wine. It's rosé." The green notes of the younger, easier-drinking braucol were now spicy and the fruit now had elegance, but it was still a wine with a wild side. The 2007 Florentin, in particular, was amazing: a Serious Wine that still had hints of braucol's bloody, ferrous edge.

After our tasting, Hirissou asked me if I'd like to meet his eagle, named Cinto. "I'm a falconer," he explained. He went outside, then moments later reappeared with an enormous golden eagle perched on his forearm, huge talons gripping a thick leather arm protector. The eagle's beak was razor sharp, but his head was covered with a tiny leather mask, so he didn't get spooked. Cinto screeched loudly, flapping his wide, powerful wings. Hirissou pulled on a small rope to keep him on his arm. "I hunt deer with him in the vineyard," he said. "He's my best friend. My wife always tells me I spend too much time with him." I was utterly speechless. In my mind, well-aged braucol will now always be inextricably wrapped up in my face-to-face encounter with a live golden eagle.

Hirissou returned his eagle outside to wherever he keeps him. His wife, Bénédicte, broke out her acoustic guitar and sang whispery French folk songs as we all sipped some more aged braucol. I showed Matteo how to swing a plastic baseball bat that his father had brought home from a sales trip to New York. Hirissou told me that he loved NBA basketball, and I told him that if he ever came to Philadelphia, I'd take him to a 76ers basketball game.

Less than a month after I'd returned home, Hirissou emailed to tell me he was coming to New York, and planned to take a train down to Philadelphia. I told him I'd get the Sixers tickets.

Hirissou accompanied me that afternoon to a staff class I was teaching at Root wine bar in Fishtown. He poured his sparkling mauzac, his white blend with loin de l'oeil, and his rouge blend with duras for the staff—who were still learning the basics. Mauzac? Loin de l'oeil? Duras? A place called *Guy-yak*? I could see a panicked look in their eyes that said, "WTF? Is this going to be on the quiz?"

After the class at Root, we went to my house in the suburbs, where Hirissou met my son Sander, now a freshman in high school—years at this point since I'd quizzed him on Le Nez du Vin. Sander was very impressed with Hirissou's video of his eagle hunting a deer. I told Hirissou that Sander was taking French class, and he asked Sander, in French, if he played any sports. Sander, who is on the crew team, said: *"Je joue le rowing."*

"Ah," Hirissou corrected, "you said, 'I play rowing.' You should say, *je fais*, I do."

At that, I laughed out loud. The Gaillac winemaker, who grows grapes that no one's heard of and who hunts with an eagle, was in my suburban living room, correcting my kid's ungrammatical French. My journey had clearly taken some strange turns.

Hirissou and I went to the Sixers game. Anyone who knows anything about NBA basketball knows that the Sixers have been bad, historically awful actually, for numerous seasons. In fact, they've openly been tanking season after season, losing on purpose in an attempt to gain higher draft picks. The Sixers organization's strategy was to stop limping along as a mediocre-to-good-enough team that might sneak into the playoffs as bottom seed, but never win anything. Instead, the Sixers tore their team apart, trading away their veterans in an attempt to stockpile younger players that one day might become superstars, and create a dominating team—essentially, go big or go home. Someday in the future, the Sixers tell us, once the team acquires enough talent, championships will happen. "Trust the Process," they say.

When I think about The Process, I think about winemakers like Giulio Moriondo in Valle d'Aosta ripping out his pinot noir to plant cornalin, petit rouge, and vien de nus. Or Nicolas Gonin in Isère tearing out his chardonnay to plant altesse, verdesse, and persan. Or Johannes Gebeshuber in Gumpoldskirchen staking his livelihood on zierfandler and rotgipfler. Or the winemakers of Tramin trying to make people fall in love with gewürztraminer. Or even Paul Hobbs, who's had huge success with Napa cabernet sauvignon, Russian River pinot noir, and Argentine malbec—yet who's now trying to sell wine made from areni that's grown in Armenia. I think about all the grapes in the world we've never yet tasted—the still-to-be-discovered ones grown by some unknown winemaker in some obscure vineyard.

Hirissou and I settled into our seats. No wine. Tonight we would drink watery Goose Island IPAs in plastic cups. By halftime, the Sixers were being trounced, as usual. Still, at one point, a familiar, if ironic, chant emerged through the stadium: "Trust The Process!" It got louder, "Trust The Process!"

"What are they saying?" Hirissou asked. I didn't even know where to begin. I tried to explain that the fans were sort of joking. That the Sixers were at the beginning of a long rebuilding process to go from last place to a winning team. That the fans were restless, but patient, waiting for the young players to develop. All in hopes that one day we will win again.

Hirissou, a man who makes wine from incredibly obscure grapes, whose region was kept down for five centuries by a behemoth like Bordeaux, seemed to intuitively understand The Process. "Yes," he said. "I think this is like wines from Gaillac. Maybe one day, in America, everyone will be drinking braucol."

We joined the chanting crowd: "Trust The Process!"

Appendix: Gazetteer of *Godforsaken Grapes*

101 Varieties to Seek Out

What follows is a glossary and pronunciation guide for the many grapes mentioned in this book. One hundred and one varieties may seem like a lot—especially since many of these will be unfamiliar to the average wine drinker, but 101 varieties is merely a drop in the ocean, a grain of sand in the desert, a blip in the universe of wine grapes. Consider that there are 1,368 known wine grapes; the ones listed here represent less than 8 percent of that total.

Full disclosure: I am not a linguist. I have, however, tried to give readers who are unfamiliar with these grapes a very basic, very American primer to help with pronunciations. (I'm certain saying this will not stop the language police's nitpicking. But, hey, I tried.)

The reason I want to help you to pronounce the names of these varieties is simple: I hope that you'll seek out the ones that interest you. That's why I'm also suggesting producers to look for. Do yourself a favor and introduce yourself to a new world of wine.

Of course, as time goes on, some of the varieties that now seem obscure will grow popular and become old hat. Some of the grapes that seem trendy now will disappear, only to be *rediscovered* decades from now by a new generation. If I revise or update this book five or ten or 20 years from now, this gazetteer will surely be different. The world of wine is a dynamic place. Each glass is only a snapshot in time.

Altesse. *ahl-TESS.* White grape found in the French Alpine regions of Isère and Savoie. Main grape of the Roussette de Savoie AOC. Producers to look for: Nicolas Gonin, Louis Magnin, and Charles Gonnet.

Amigne. *ah-MEEN-yeh.* Ancient white grape grown in Valais, Switzerland, particularly around Vétroz. Mostly used for sweet or off-dry wines, but look for dry bottlings by Domaine Jean-René Germanier.

Antão Vaz. *an-TOW vajsh.* White variety from Portugal's Alentejo region. Generally found in blends, such as those from Alentejo producer Esporão.

Areni. *ah-REH-nee.* The major red grape of Armenia, and possibly one of the world's oldest wine grapes. Bears the same name as a cave complex where archaeologists found evidence of the earliest known winemaking, dating to more than 6,000 years ago. Producers: Yacoubian-Hobbs, Zorah, Trinity Canyon, and Koor.

Arinto. *ah-RIHN-toh.* White variety from Portugal, also known as pedernã in the north. It can be found in blends from regions such as Bucelas (near Lisbon), the Alentejo, and Minho, where it's used in vinho verde.

Arvine. *ar-VEEN.* Also known as petite arvine; white grape grown in Valais, Switzerland, and Valle d'Aosta, Italy. Producers: Domaine Jean-René Germanier, Marie-Thérèse Chappaz (Switzerland), and Grosjean Freres (Italy).

Baco noir. *BAH-koh nwahr.* Red hybrid of European and North American grape species. Used for wine that's distilled into Armagnac brandy.

Baga. *BAH-guh.* Spicy, savory red grape from Bairrada, Portugal. Look for producers such as Sidonio De Sousa or Filipa Pato.

Blauburger. *blouw-bur-ger.* Red grape from Austria; cross between blaufränkisch and blauer Portugieser created by Dr. Fritz Zweigelt. Very obscure, most likely found by the glass in a local Viennese heuriger.

Blaufränkisch. *blouw-FRANN-kish.* Red grape grown mainly in Austria, but also Germany and New York's Finger Lakes, where it's called lemberger, and Hungary, where it's called kékfrankos.

Boğazkere. *bow-aahz-keh-reh.* Turkish for "throat burner," this tannic, savory red is from southeastern Anatolia, Turkey. Look for bottlings from Kayra or Kavaklidere.

Brianna. *bree-AH-nah.* White hybrid of European and North American grape species, and grown mostly in the upper Midwest. Look for La Garagista's lightly sparkling wine *pétillant naturel* Ci Confonde produced in Vermont.

Cot. *caht.* Most wine drinkers know malbec, but the red grape originated in Cahors, France, where it's called cot and they've been producing big, powerful wines. Producers: Château Haut-Monplaisir, Château du Cèdre, and Simon Busser's "Pur Cot."

Castelão. *cast-ehl-OW.* One of the most widely grown reds in Portugal, and sometimes used in Port blends. The main grape of the affordable, widely available Periquita wines.

Chambourcin. *shahm-boor-SAN.* A hybrid of French and North American grapes, created in the 1860s in France. It's widely grown in the United States, mostly in states such as Virginia, Pennsylvania, Missouri—and especially New Jersey, from producers such as Heritage and Sharrott.

Chasselas. *shahs-suh-LAH.* Switzerland's most-planted white grape. Known as fendant in the Swiss region of Valais.

Croatina. *crow-ah-teena.* Often called bonarda in northern Italy, this red grape is blended with barbera to create the Gutturnio wines of Colli Piacentini. Look for Gutturnio from Torre Fornello, Podere Casale, or Montesissa.

Diolinoir. *dee-oh-lee-nwahr.* Rare Swiss red variety, created in 1970 from a cross between pinot noir and a grape called robin noir. Producers: Domaine de Beudon, Domaine Jean-René Germanier, and Robert Gilliard.

Duras. *dew-RAH.* Red grape grown mainly in Gaillac, in southwest France. Takes its name from the French word for "hard," because of its hard rootstock, and how difficult it can be to grow. Producers: Domaine Plageoles, Domaine du Moulin, and Domaine des Terrisses, L'Enclos des Braves.

Emir. *eh-MERE.* This lively white is native to the historic Cappadocia region of Turkey, where it's believed winemaking dates back more than 7,000 years. Look for producers such as Turasan and Kayra.

Encruzado. *ehn-croo-ZAH-do.* Prized white variety from Portugal's Dão region. Producers: Casa de Mouraz, Quinta do Perdigão, Álvaro Castro, Quinta de Cabriz.

Étraire de la dhuy. *AYE-treyr day la DOO-ee.* Very rare red Alpine variety grown in Isère and Savoie. Producers: Domaine Finot and Domaine des Rutissons.

Fer servadou. *fair SAIR-vah-do.* Savory red grape grown in southwest France, where it is known as braucol in Gaillac and mansois in Marcillac. Producers: Domaine du Moulin (Gaillac), Domaine des Terrisses (Gaillac), and Domaine du Cros (Marcillac).

Friulano. *free-oo-LAH-noh.* One of the main white grapes of Friuli, the northeast region of Italy bordering Slovenia. Formerly called tocai, it's called ravan in Slovenia, and sauvignonasse (or sauvignon vert) in southwest France, where it's believed to have originated.

Fumin. *foo-MAHN.* Red grape from Valle d'Aosta, Italy, often blended with petit rouge and cornalin. Producers: Grosjean Freres, Les Crêtes, Lo Triolet.

Furmint. *FOOR-mint.* One of the major white grapes of Hungary, historically blended in the prized Tokaji sweet wines. Now, it's often bottled as a single variety, and dry.

Frontenac. *FRON-ten-ack.* North American hybrid grape created from a wild grapevine in Minnesota crossed with at least eight different grape species. Mainly grown in the northern US; La Garagista, in Vermont, uses the grape in its blends.

Frühroter veltliner. *froo-ROH-ter VELT-lee-ner.* White variety found in blends from Austria's Thermenregion. Means "early red" veltliner, in contrast to zierfandler, which is nicknamed "late red," or *spaetrot.*

Gelber muskateller. *GEHL-ber MOOSK-ah-tell-er.* Also known as muscat blanc à petits grains. Aromatic white grape grown in Austria and elsewhere. Producers: Hermann Moser, Heidi Schröck, Berger, and Knoll.

Gewürztraminer. *guh-vurtz-TRAH-mee-ner.* Aromatic white grape grown in Alto Adige, Alsace, California's Anderson Valley, and elsewhere. Producers: Cantina Tramin, Elena Walch (Alto Adige), Trimbach, Hugel, Domaine Zind-Humbrecht (Alsace), Husch, Handley, Phillips Hill, and Navarro Vineyards (Anderson Valley).

Godello. *goh-DAY-yoh*. White grape mainly grown in Galicia, in northwest Spain, in the regions of Valdeorras, Ribeira Sacra, and Bierzo. Nearly extinct as recently as the 1970s, it now produces some of Spain's most popular whites.

Goldburger. *gold-bur-ger*. Austrian white grape that is a cross between welschriesling and orangetraube. Often in Viennese gemischter satz blends.

Gouais blanc. *goo-WAY blahnk*. Called gwäss or heunisch in German. This white grape has been banned across Europe, by various royal decrees, since the Middle Ages. Monarchs considered it a prolific, peasant grape that made bad wine—*gou*, in medieval French, along with *Heunisch* were derogatory terms for low quality. Through DNA testing, however, gouais blanc was found to be the mother of around 80 varieties, several with pinot noir as the father, including chardonnay, gamay, and possibly riesling.

Grignolino. *gree-nyoh-LEE-noh*. Red grape from Italy's Piedmont region, particularly Asti and Casale Monferrato, that produces light ruby wines with high acidity and significant tannins.

Gringet. *grahn-JZEY*. Alpine white grape from Savoie region, grown on less than 100 acres. Most notable from the vineyards of Ayse, in particular by Domaine Belluard.

Gros manseng. *grow MAHN-song*. Aromatic white grape from southwest France, most notably from Jurançon and Gascony, often found in blends with petit manseng, sauvignon blanc, and others.

Grüner veltliner. *GREW-ner VELT-lee-ner*. Major, indigenous white variety of Austria with a wide range of styles. Noteworthy regions include Wachau, Kamptal, and Kremstal.

Hárslevelü. *HARSH-leh-veh-LOO*. One of the major white grapes of Hungary, historically blended in the prized sweet wines of Tokaji.

Hondarrabi zuri. *on-da-rabb-eh zorr-ee*. White grape from the Basque region of Spain used to make txakolí (*CHA-ko-lee*).

Humagne blanche. *hoo-MAN-yeh blahnsh*. Very obscure Swiss white grape grown on less than 100 acres, mostly in Valais. Not related to humagne rouge.

Humagne rouge. *hoo-MAN-yeh rooj.* Alpine red variety, grown mainly in Switzerland and also in Valle d'Aosta, Italy, where it's known as cornalin.

Jacquère. *jah-KEHR.* Most widely planted white grape in France's Savoie region, and best known from the crus of Apremont and Abymes. Producers: Charles Gonnet, Gilles Berlioz.

Juhfark. *YOO-fark.* Hungarian white grape whose name means "sheep's tail" because of the elongated, curved shape of the clusters. Most significantly grown in the volcanic soils of Somló by producers such as Fekete Béla.

Kadarka. *kah-DARK-ah.* The best-known Hungarian red grape, from which the famed Bikavér, or "Bull's Blood," of Eger is made. It's also grown in Serbia, Bulgaria, and Romania. Producers such as Eszterbauer and Heimann make light-bodied, spicy, eminently drinkable expressions.

Kerner. *KEHR-ner.* White grape, a cross between riesling and schiava, that's grown mainly in Alto Adige / Südtirol and Germany. Look for Abbazia di Novacella's single-varietal bottling.

Kisi. *KEE-see.* Unusual white grape, grown on less than 125 acres in the Republic of Georgia, which claims to be the birthplace of winemaking. Doqi produces a bottling that is aged in traditonal qvevri.

La crescent. *lah KREH-sent.* North American hybrid that's a cross between muscat and an unidentified native grape from the *vitis riparia* species, created at the University of Minnesota in the 1980s. Grown in the Midwest and northern United States.

Lagrein. *lah-GRAH'EEN.* Full-bodied red grape grown mainly in Alto Adige / Südtirol. Producers: Tiefenbrunner, Castel Sallegg, and Muri-Gries.

Listán negro. *lee-STAN NEY-grow.* Grown on the volcanic soils of the Canary Islands, this grape produces light reds.

Len de l'el (also spelled loin de l'oeil). *len-deh-LEYHL.* White grape from southwest France, particularly Gaillac, which is a complete mystery to ampelographers. Producers: Domaine du Moulin, Domaine des Terrisses, and L'Enclos des Braves.

Marquette. *mar-KETT.* North American hybrid created at the University of Minnesota. Grandparent is pinot noir, but is made up of several species such as *vitis riparia, vitis labrusca, vitis rupestris*, and others. Grown in midwestern and northern US states, including by La Garagista in Vermont.

Mauzac. *mohw-ZACK.* White variety often used in the sparkling wines of southwest France, in particularly Gaillac and Limoux, which historically predate the sparkling wines of even Champagne. Producers: Domaine du Moulin, Domaine Plageoles (Gaillac), and Domaine de Martinolles (Limoux).

Mencía. *mehn-THEE-ah.* Iberian red grape grown mainly in Spain's Bierzo, Ribeira Sacra, and Valdeorras regions. It's also grown in Portugal, where it's known as jaen.

Mondeuse. *mohn-DOOZ.* Red variety grown in the French Alps, most notably Savoie, with the finest expressions coming from around the village of Arbin. Producers: Domaine Prieuré Saint Christophe, Domaine Louis Magnin, and Maison Philippe Grisard.

Müller-thurgau. *MEW-luhr TOOR-gow.* White variety grown in Germany, Austria, Switzerland, Alto Adige / Südtirol, and elsewhere. Cross between riesling and madeleine royale, created in the late 19th century by Dr. Hermann Müller of Thurgau, Switzerland. Tiefenbrunner's Feldmarschall von Fenner bottling is especially good.

Négrette. *neh-GREHT.* Main red grape of Fronton AOC, near Toulouse in southwest France, from which bold, aromatic reds and refreshing rosés are made. Producers: Château Bellevue La Forêt, Domaine Roumagnac.

Neuburger. *NOY-burger.* Austrian aromatic white grape that's a cross between sylvaner and roter veltliner. Notably grown in the Wachau and Burgenland, especially Leithaberg DAC.

Ondenc. *ON-dank.* White variety from southwest France, particuarly in Gaillac, that almost disappeared in the 20th century. Domaine Plageoles and others have recently revived the grape.

Orangetraube. *ORANGE-trowb.* Not to be confused with so-called "orange wine." Austrian white grape so obscure that it's not an official grape under the *qualitatswein* classification. Zahel produces an excellent single-varietal bottling called Orange T.

Ortrugo. *OR-troo-go.* White grape found in the Colli Piacentini DOC of Italy's Emilia-Romagna. Often blended with malvasia and sometimes used in frizzante or sparkling wines.

Persan. *per-SAHN.* Alpine red variety grown in Savoie and Isère. In the 18th century, it was considered one of the best red wines in France, but now there are only about 25 acres in existence. Look for bottlings by Nicolas Gonin and Domaine Giachino.

Petit manseng. *peh-TEE MAHN-song.* Aromatic white grape from southwest France, most notably from Jurançon and Gascony, often found in blends with gros manseng, sauvignon blanc, and others.

Petit rouge. *peh-TEE rooj.* Red grape from Valle d'Aosta, Italy. Main grape of subregion Tourette, which must have 70 percent. Producers: Grosjean Freres, ViniRari, and Maison Anselmet.

Petite sirah. *peh-TEET sih-RAH.* Originally called durif (*doo-REEF*) and originally found in the French Alps, before it migrated to California. Now, under its new Americanized name, it's best known for its deep, dark reds, and for being a blending partner with zinfandel.

Petit verdot. *peh-TEE vehr-DOH.* One of the grapes permitted in the classic red blends of Bordeaux, where it's used in very small amounts. Also grown in Australia and the United States.

Prié blanc. *pree-EH blahnk.* White grape from Valle d'Aosta, Italy. Main grape of subregion Morgex et de La Salle, near Mont Blanc. Producers: Ermes Pavese, Cave du Vin Blanc.

Prunelart. *proo-neh-LAHR.* Forgotten red variety from Gaillac, in southwest France, known for its floral aromas. As recently as 1998, there were fewer than five acres. Has been allowed for production in the Gaillac AOC only since 2017. Producers: Domaine Plageoles, Domaine des Terrisses.

Ramisco. *rah-MEESH-koh.* Red grape grown almost exclusively in the Portuguese region of Colares, where it survived phylloxera in the 19th century. Producers: Adega Regional de Colares, and Adega Viúva Gomes.

Räuschling. *ROWSH-ling.* Rare white grape grown around Lake Zürich. Most likely found poured in a Zürich wine bar.

Refosco. *Reh-FOHS-koh.* Ancient red variety of Friuli, in northeast Italy. Most famous variation is refosco dal peduncolo, of which the best expression is the Colli Orientali, from producers such as Livio Felluga, Bastianich, and La Viarte.

Ribolla gialla. *ree-BOHL-lah JAHL-lah.* An ancient variety and one of Friuli's major white grapes. It's also grown across in Slovenia, where it's called rebula. Used in the popular orange wines by Friuli winemaker Josko Gravner.

Rkatsiteli. *ruh-KAT-see-TELL-ee.* Most widely planted white grape in the Republic of Georgia, where it is often used in traditional qvevri wines, such as those by Doqi and Pheasant's Tears. It's also grown in New York's Finger Lakes by Dr. Konstantin Frank.

Roter veltliner. *ROH-ter VELT-lee-ner.* "Red" veltliner, though it's actually not related to grüner veltliner. Parent of Austrian varieties neuburger, rotgipfler, and zierfandler. Finest expression comes from the Wagram region, from winemakers such as Franz Leth and Anton Bauer.

Rotgipfler. *ROHT-gihp-fluhr.* White wine from Austria's Thermenregion, where it was traditionally blended with zierfandler in the famed wines of Gumpoldskirchen. Look for the excellent examples by Weingut Gebeshuber.

Roussanne. *roo-SAHN.* Best known as a variety from Rhône, where it's often blended with marsanne. However, roussanne finds one of its most fascinating expressions in Savoie, in Chignin-Bergeron. Producers: Gilles Berlioz, Adrien Berlioz, and Louis Magnin.

Sagrantino. *SAH-grann-TEE-noh.* Red grape of Montefalco, in Italy's Umbria region, where big, tannic reds are produced. Producers: Arnaldo Caprai, Paolo Bea, Antonelli, Scacciadiavoli, and Còlpetrone.

Saperavi. *SAH-per-ah-vee.* Main red grape of the Republic of Georgia. Name means "dye," and it's one of the few grape varieties in the world where both the skin and the juice inside have a red pigment. Look for bottlings by Doqi and Pheasant's Tears from Georgia, as well as from Dr. Konstantin Frank in New York's Finger Lakes.

Sankt Laurent. *just say "Saint Laurent."* Austria's finicky, pinot-like red grape. Producers: Juris, Erich Sattler, Umathum, Bründlmayer.

Savagnin. *sah-vah-NYAHN*. Also known as heida (Switzerland) and traminer (throughout the German-speaking Alps). In the French region of Jura, it's known for making vin jaune, the sherry-like "yellow wine."

Scheurebe. *SHOY-reyb-beh*. Lush, aromatic white grape grown in Germany and Austria, where it's called sämling 88 (seedling 88). Importer Terry Theise describes it as "like riesling just after it read the *Kama Sutra*."

Schiava. *ski-AH-vah*. Red grape from Alto Adige / Südtirol (where it is also known as vernatsch) and also Germany, where it is called trollinger. Creates bright, light-bodied, almost-pink red wines. In Alto Adige / Südtirol, it's used, along with lagrein, in one of the best-known reds of the region, Santa Maddalena DOC.

Schioppettino. *skyow-peh-TEE-noh*. Red grape found mainly in Italy's Friuli region, where it is sometimes called ribolla nera, though it is not related to ribolla gialla. Dates to the 13th century, but almost disappeared by the late 20th century until being revived along the Friuli-Slovenian border.

Sercial. *SER-shuhl*. The driest of the grapes grown on Madeira to make the island's famous fortified wines.

Sémillon. *seh-mee-YOWN*. One of the grapes permitted in the classic white blends of Bordeaux, particularly the sweet wines of Sauternes. Also famously grows in Australia's Hunter Valley, as well as the United States.

Silvaner. *sihl-VAN-uhr*. White grape grown primarily in Germany (particurlarly Franconia and Rheinhessen) and Alsace (where it's spelled sylvaner). Originated in Austria, though there is little left there. Gained a bad reputation in the 1970s from low-quality bulk *Liebfraumilch* wines, such as Blue Nun, but now is regaining its stature, particularly from a new generation of producers in Rheinhessen.

Tannat. *tuh-NAHT*. Grown in southwest France, most notably in Madiran, and also some in Uruguay. Scientists have found it contains the highest, most potent levels of polyphenols, those antioxidants that prevent an array of health problems like cancer, heart disease, and diabetes.

Teroldego. *teh-ROHL-deh-goh*. Red grape grown mostly in Italy's northern Trentino region. Look for the Teroldego Rotaliano DOC and producers such as Foradori.

Timorasso. *tee-MORE-ah-so.* After phylloxera, timorasso had dwindled to less than 20 acres around the town of Tortona, in the Piemontese province of Alessandria. In the early 1990s, a winemaker in Tortona named Walter Massa nurtured the grape back from the brink of extinction.

Tintilla. *tin-TEE-ah.* Local name in Spain's Andalusia region for the red grape also known as graciano. While used in Rioja as a blending grape with tempranillo, in Andalusia there are single-varietal bottlings by producers such as Vara y Pulgar.

Touriga fêmea. *too-REE-guh feh-MAY-eh.* Incredibly rare red variety from Douro, Portugal. Look for Quinta da Revolta's single-varietal bottling.

Touriga nacional. *too-REE-guh nah-syoo-NAHL.* One of the major red grapes of Portugal, often used in dry reds from Douro. Historically among the grape blends used to make Port, along with touriga franca, touriga francisca, tinto cão, tinta roriz (aka tempranillo), and others.

Trousseau. *TROO-soh.* Known as bastardo in Portugal (where it can be used in Port wine). It's the most coveted red grape from the French region of Jura.

Turbiana. *TOOR-bee-anna.* The Lombardian name for verdicchio grown in Lugana, on Lake Garda. A quarter of Lugana's vines are threatened by the proposed construction of a high-speed train. #SaveLugana. Producers: Zenato, Tenuta Roveglia, and Cà dei Frati.

Verdesse. *VEHR-dess.* Relativelty obscure, aromatic Alpine white that grows mainly in Savoie and Isère. Producers: Nicolas Gonin, Domaine Finot.

Vien de nus. *vee-EN de NOOS.* Red grape grown in Valle d'Aosta, Italy, generally in blends with petit rouge, fumin, or cornalin by producers such as Grosjean Freres or ViniRari.

Voskehat. *voh-ski-hot.* Meaning "gold berry," this is the most important white variety of Armenia, notably from the Aragatsotn region, from producers such as Koor.

Welschriesling. *VELSH-reez-ling.* Not related to riesling, this white is grown in Austria and Germany, as well as Hungary, where it's known as olaszrizling, and Croatia, where it's known as graševina. Producers: Heidi Schröck (Austria), Fekete Béla (Hungary), and Adžić (Croatia).

Xinomavro. *ksee-NOH-mah-vroh.* Literally means "acid black," this red grape—often compared to Italian nebbiolo—is grown mostly in the Naoussa region of Macedonia in northern Greece. Producers: Domaine Karydas, Kir-Yianni, and Boutari.

Zierfandler. *zeer-FAND-ler.* White wine from Austria's Thermenregion, where it is traditionally blended with rotgipfler in the famed wines of Gumpoldskirchen. Also known as spaetrot, or "late red," because it turns red in the fall before harvest (even though it produces a white wine). Look for the excellent examples by Weingut Gebeshuber.

Žilavka. *zhe-LAHV-ka.* White grape of Bosnia and Herzegovina, where it grows primarily in the Mostar region.

Zweigelt. *TSVY–gelt.* The most widely planted red grape in Austria. Also known as rotburger, it's a cross between blaufränkisch and sankt laurent.

Acknowledgments

I AM INDEBTED TO JOSÉ VOUILLAMOZ, Julia Harding, and Jancis Robinson for their comprehensive, beautiful, and defining work on ampelography, *Wine Grapes: A Complete Guide to 1,368 Vine Varieties, Including Their Origins and Flavors*. This 1,242-page tome served as an invaluable reference, as well as an endless source of inspiration while I was writing *Godforsaken Grapes*. What's perhaps most inspiring is that the work of discovering and rediscovering wine grapes, decoding their origins, and explaining their complexities is never-ending—Vouillamoz tells me that so much new research has happened since the book's original publication. I eagerly await the trio's future, updated editions.

I would also like to acknowledge the support of all my families, in New Jersey (Jen, Wes, and Sander), Florida (Frank and Becky), California (Jack and Mariann), and Pieve San Giacomo (Anna and Daniela). Many thanks to Jamison Stoltz at Abrams for taking a chance on this strange book, and for learning a little about wine along the way. Thanks also to Derk Richardson and Jeremy Saum at *AFAR*, David Rowell and Joe Yonan at the *Washington Post*, and William Tish at *Beverage Media*, who all helped usher pieces of this book into print. Many thanks to Constance Chamberlain, the Inama family (Stefano, Matteo, Alessio), David Foss, and the people at Wine Mosaic, who have been true wine-biz friends, as well as Greg Root and the original Fishtown Wine Club (Lauren, Joe, Mike, Alexa, Francisco, Nicole, and Amber). Thanks to my brother Tyler for the long-ago drama of the Bedford Rascal, to Kevin Dorn for his brilliant headshots (he had little to work with), and Shelby Vittek, irrepressible redhead, for her inspiring journey from Franzia-guzzling novice to wine expert. Above all, thanks to all the independent-minded, potentially crazy winemakers around the world who are committed to growing grapes you've never heard of.

Index